Salah-Eddin Gherbi

Scala Harmonica

The Geometry of Planetary Resonance

Vektor Publishing

First edition, 2026

Website: `https://salaheddingherbiauthor.com/`

ORCID: `https://orcid.org/0009-0005-4017-1095`

Substack: `https://salaheddin.substack.com/`

Research: `https://doi.org/10.5281/zenodo.18816002`

ISBN 978-1-83709-520-9

Published by Vektor Publishing

Preface

The solar system has always interested people, not only because of its beauty but also its mysteries. However, it may also show a hidden order. Planets move around the Sun. There may be small mathematical connections that link planet positions. These links may be like patterns in sacred geometry.

This book looks at one geometric shape: the Celtic Cross. Its concentric circles and intersections generate ratios based on the square root of 2 and related numbers. Examples of these numbers are $\sqrt{2}+1$, $2\sqrt{2}-1$, and $\sqrt{2}-1$, which are basic to the design. They help us look for possible cosmic alignments. By studying these ratios and their results, we can build harmonic ladders. These ladders suggest interesting links between planet distances.

The main idea is straightforward yet profound. What if the distance between planets in our solar system is not just by chance? What if it shows a harmonic pattern? For example, Venus's distance from the Sun follows a silver-ratio relationship ($\frac{1}{\sqrt{2}} \approx 0.707$), which is remarkably close to its actual orbital radius of 0.722 AU. Additionally, a lost planet between Mars and Jupiter—which we shall call *Harmonia*— may undergo a very close alignment, as clearly shown in the harmonic ladder that leads to Jupiter. Planets farther out also follow ratios that reflect these geometric rules.

This study uses tables, diagrams, and harmonic ladders. It visually displays these links, allowing the reader to see patterns that numbers alone might conceal. The aim is not to give final scientific proof. The purpose is to show the language of ratios and geometry. This language has interested mystics, mathematicians, and astronomers.

In the following pages, we study, compare, and place each planet interval within the Celtic Cross geometry. The diagrams connect abstract mathematical concepts with the fundamental structure of our solar system. They ask the reader to consider the idea that numbers and shapes may represent the cosmic order.

Contents

I

Celtic Cross Geometry

The Celtic Cross is a notable geometric shape. Complex balanced measurements come from a simple repeating rule. The design starts with one square, copying itself in eight directions — up, down, left, right, and diagonally — forming a 3×3 grid.

Fig. 1. The Celtic Cross emerges from an eight-pointed star drawn with a compass and a straightedge.

Geometric Construction

Building the Foundation

Draw vertical and horizontal center lines on the first square. Also, draw a circle around it. These parts help to add another square of the same size. This new square sits in the middle but turns diagonally.

Extend the lines of these two squares until they first cross. An eight-pointed star then appears. This step forms the first part of the Celtic Cross (Fig. 2).

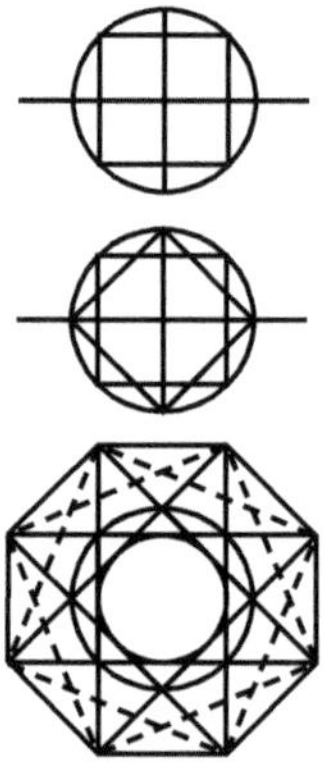

Fig. 2. First stage of the Celtic Cross geometric construction.

Completing the Cross

The centers of the four small circles coincide with the four corners of the first square. Each circle touches the circle drawn inside that square. Next, four more circles are part of the design. They have the same radius as the center circle. Their centers sit on the diagonal lines. A square formed around these four new circles completes a regular 3×3 grid, thereby forming the base of the Celtic Cross. We can see this in past examples, such as the back of the Rosemarkie Stone (Fig. 3).

Fig. 3. The back of the Rosemarkie Stone.

Harmonic Ratios Revealed

Emergent Proportions

This geometric setup naturally shows interesting ratios between the circles:

- The central circle (inside the eight-pointed star) relates to the circle going through the centers of the four small corner circles as $1 : \sqrt{2}$ (Fig. 5). This ratio also matches the distance from fingertips to the deltoid tuberosity of the humerus (Fig. 4). This distance defines the Egyptian Royal Cubit as a sacred proportion in the human body. Each small corner circle has a diameter of $\sqrt{2} - 1$ when the central circle's diameter is 1 (Fig. 5).

Human and Cosmic Connections

These ratios are not just for art. They hold a harmonic structure that we explore in cosmic settings. By changing these ratios into numbers, we build ladders and comparisons. These show possible links with planetary semi-major axes and other natural harmonic systems. The Celtic Cross serves as a visual and numerical guide, connecting geometric shapes, human proportions, and cosmic distances.

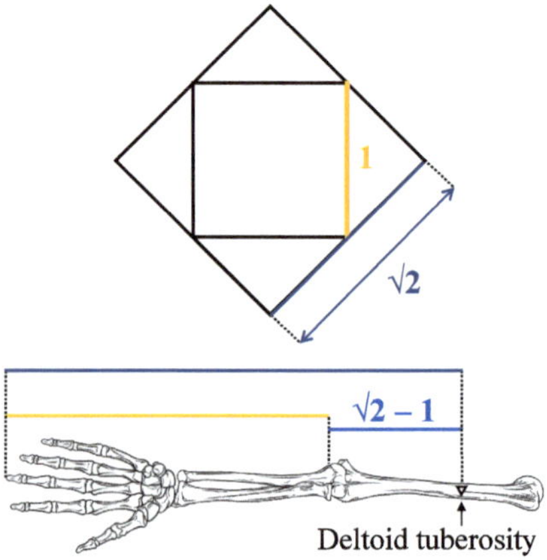

Fig. 4. The ratio $1 : \sqrt{2}$ determines the arm's proportion, and it also appears in nested squares.

The Royal Cubit measures the forearm's ratio to the humerus's deltoid tuberosity ($1 : \sqrt{2}$), a fundamental proportion within the human body. This triangular area on the anterolateral (front side) lies proximally about a third of the way to the elbow. It serves as an attachment point for the deltoid muscle.

Core Ratios and Their Values

The Fundamental Constants

The Celtic Cross geometry shows a group of harmonic values based on $\sqrt{2}$. These ratios come from the concentric circles and intersections in the design. They form the basis for possible cosmic links (Fig. 5).

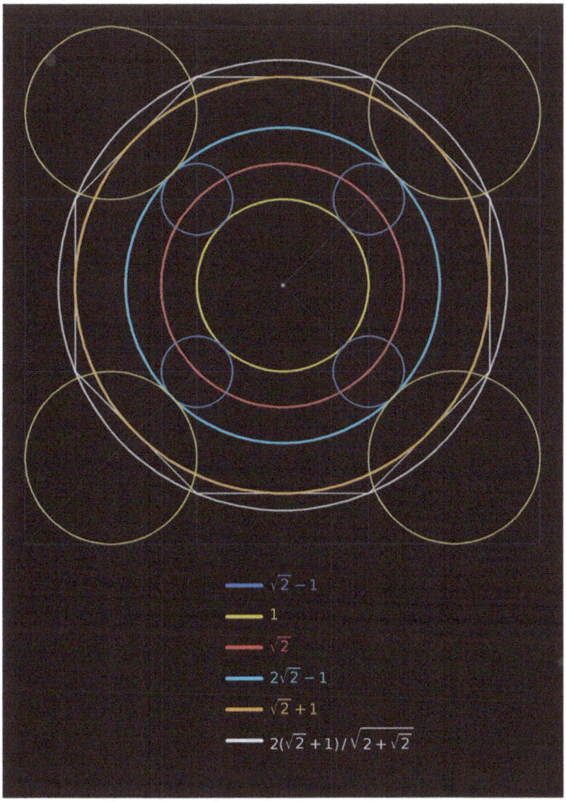

Fig. 5. Concentric Circles Illustrating Celtic Cross Ratios.

The **core ratios** are:

- $A = \sqrt{2} \approx 1.414$
- $B = \sqrt{2} + 1 \approx 2.414$
- $C = 2\sqrt{2} - 1 \approx 1.828$
- $D = \sqrt{2} - 1 \approx 0.414$

Derived Relationships

Each of these values is a scaling factor, similar to climbing a harmonic ladder. When we examine the planets' orbital structures, some fascinating similarities begin to emerge.

Combining these values by division creates more ratios:

- $C/D \approx 4.414$
- $B/A \approx 1.707$
- $C/A \approx 1.293$
- $A/D \approx 3.414$

Table 1 lists the core and derived Celtic ratios along with their mathematical expressions and decimal values rounded to three decimal places.

Table 1. Core and derived Celtic ratios (decimal approximations).

Label	Expression	Value
A	$\sqrt{2}$	1.414
B	$\sqrt{2}+1$	2.414
C	$2\sqrt{2}-1$	1.828
D	$\sqrt{2}-1$	0.414
C/D	$(2\sqrt{2}-1)/(\sqrt{2}-1)$	4.414
B/A	$(\sqrt{2}+1)/\sqrt{2}$	1.707
C/A	$(2\sqrt{2}-1)/\sqrt{2}$	1.293
A/D	$\sqrt{2}/(\sqrt{2}-1)$	3.414

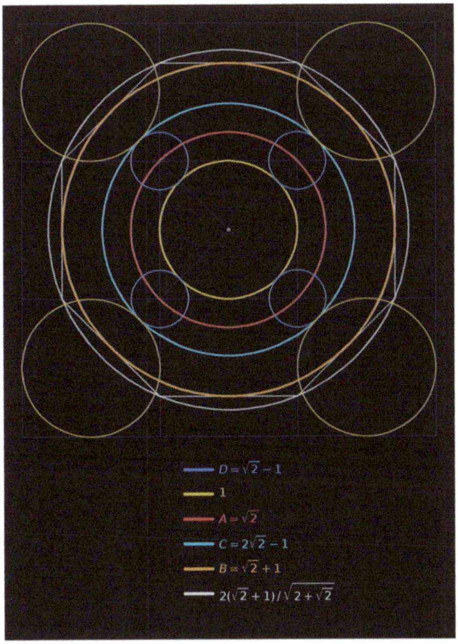

Fig. 6. Concentric Circles Illustrating Celtic Cross Ratios with their corresponding letters A, B, C and D.

From Geometry to Cosmos

Planetary Correlations

These other relationships stand out when compared to orbital distances or periods. For example:

- The ratio $C/D \approx 4.414$ is close to the spacing between Saturn and the theoretical Harmonia node: $a_S/a_H \approx 9.559/2.142 = 4.463$ (the Harmonia value derived from Celtic Cross geometry; see Chapter 3).
- The ratio $B/A \approx 1.707$ resonates with the Venus-Earth relationship: since $1.707 = 1 + \sqrt{2}/2$, and Venus orbits at ≈ 0.722 AU (close to $\sqrt{2}/2 \approx 0.707$), this ratio captures the complementary relationship between the two planets.
- The ratio $A/D \approx 3.414$ is close to the orbital spacing of Jupiter to Mars in semi-major axes: $a_J/a_M \approx 5.204/1.524 = 3.415$.

The Harmonic Ladder

The above instances depict a harmonic ladder composed of proportions of the square of two ($\sqrt{2}$). This arrangement forms a geometrical structure of the Celtic Cross and relates to the spatial arrangement of the Solar System. It is unclear whether these associations are mere coincidences or symbolic patterns, or whether they contain any substantive connection between the geometry and the universe; therefore, we should conduct a more thorough exploration of the planetary harmonics.

Such interrelationships confirm that the ratios represented by the Celtic Cross are not merely abstract constructs; instead, they are deeply rooted in empirical reality and reflected in planetary and orbital harmonics within the natural world.

The Universal Language of $\sqrt{2}$

The prominence of $\sqrt{2}$ in the Celtic Cross is not merely artistic coincidence. This same ratio appears throughout physics wherever symmetry and duality converge—between wave and particle, rest and motion, complementary states. It emerges in quantum superposition ($\frac{1}{\sqrt{2}}(|\uparrow\rangle + |\downarrow\rangle)$), in the conversion between peak and RMS voltage, and in the Lorentz factor when kinetic energy equals rest energy. Even the brain's grid cells organize in hexagonal lattices whose scaling follows $\sqrt{2}$.

That this ratio recurs across such different domains hints at something deeper—a geometric rhythm underlying matter and energy. The Celtic Cross, then, does not stand apart from the natural sciences; it reflects harmonic principles evident throughout the natural world.

With this in mind, we now turn to the **inner planets**—Mercury, Venus, Earth, and Mars—to explore how precisely their semi-major axes resonate with these $\sqrt{2}$-derived proportions. Here, the Celtic Cross ratios emerge as more than geometry: they echo the harmonic order of planetary space.

2

Inner Planet Harmonics

THE INNER PLANETS — Mercury, Venus, Earth, and Mars — offer the first test for the Celtic Cross harmonic ladder. Their semi-major axes are known. When compared to the $\sqrt{2}$ ratios found earlier, a striking set of resonances appears. Through careful analysis, we find more precise harmonic matches. These show the mathematical order of planet spacing.

The Harmonic Pattern Revealed

Planetary Alignments

The careful harmonic analysis shows great precision:

- Mercury's orbit distance of 0.387 AU matches closely with $\frac{14}{15}D \approx 0.387$ (a finely tuned adjustment of the baseline D), yielding a difference of just -0.10%. This tuning of the fundamental D ratio suggests Mercury holds an exact harmonic position *(Fig. 8)*.

- Venus orbits at 0.722 AU, within -2.06% of $1/\sqrt{2} \approx 0.707$. Not far from a $\sqrt{2}$ proportion compared to Earth's 1.000 AU *(Fig. 9)*.
- Earth, at precisely 1.000 AU, acts as the natural point for harmonic comparison *(Fig. 10)*.
- Mars is at 1.524 AU, showing only a $+1.48\%$ difference from the ratio $(B + D)/C \approx 1.547$ AU, which is much better than the -7.2% difference from $A = \sqrt{2}$ alone *(Fig. 11)*.

Quantifying the Precision

Throughout this book, percentage deviation is defined as:

$$\text{Dev}(\%) = \frac{\text{Predicted} - \text{Observed}}{\text{Observed}} \times 100$$

A positive value means the model predicts a larger distance than observed; a negative value means it predicts smaller.
We sum up these precise matches in the table below. It measures the alignment between planet distances and Celtic Cross ratios:

Table 2. Inner planet harmonics with Celtic ratios

Planet	Axis (AU)	Celtic Ratio	Value	Dev.
Mercury	0.387	$\frac{14}{15}D$	0.3867	-0.10%
Venus	0.722	$1/A$	0.707	-2.06%
Earth	1.000	Reference	–	–
Mars	1.524	$(B+D)/C$	1.547	+1.48%

Visualizing the Harmonic Ladder

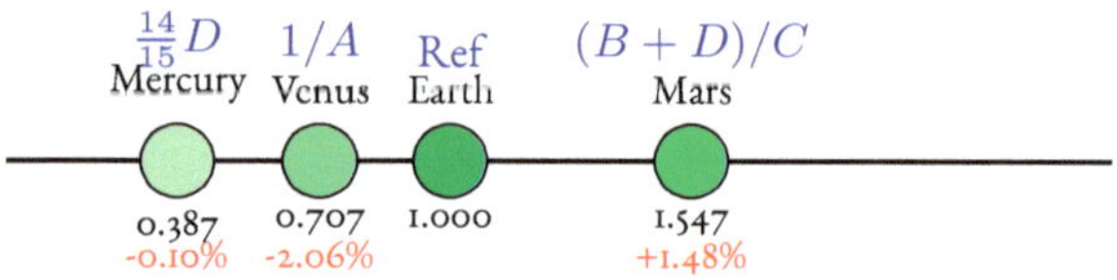

Note: Refined ratios show improved precision: Mercury as $\frac{14}{15}D$, Mars as $(B + D)/C$

Fig. 7. Inner Planet Harmonic Ladder (Earth = 1 AU).

Mathematical Foundations

Refined Ratios vs Basic Proportions

The smooth harmonic mapping exhibits outstanding mathematical harmony in the internal planetary system. The orbit of Mercury has a near-concordance with the fractional coefficient, which is, by implication, a finely tuned fractional alignment, with the basic ratio, $\frac{14}{15}D$ — an empirical adjustment of the baseline D that brings the prediction dramatically closer to Mercury's observed orbit. At the same time, the orbit of Mars (at $(B + D)/C$) demonstrates the ability of compounded ratios to give better harmonic fits compared to simple geometric constants.

The significant ratios of Compound Celtic Cross yield a significant improvement over simple geometric values: the deviation of Mars becomes less significant, of -7.2% (in comparison with $A = \sqrt{2} \approx 1.414$ AU) and a more modest $+1.48\%$ (in comparison with $(B + D)/C \approx 1.547$ AU). Similarly, the orbit of Mercury matches the value of $\frac{14}{15}D$ nearly perfectly (with 0.387 AU), but the ratio of the baseline D would add a -6.5% error.

Celtic Cross Geometry in Action

The inner planets match the square-root-of-2 ratios in the Celtic Cross. Mercury, Venus, Earth, and Mars fit positions that map onto this harmonic ladder, suggesting that the same geometric rules seen in the Celtic Cross also shape our Solar System.

Mercury, the nearest planetary body, aligns perfectly with $\frac{14}{15}D$, thus forming the first step of the harmonic ladder with practically no deviation. This fractional correspondence indicates a complex harmonic tuning system that goes beyond the simple derived ratio of the root of 2. Venus, a planet with a slightly larger heliocentric distance, is in harmony with the ratio $1/A \approx 0.707$ with only a -2.06% difference, much closer to the inverse ratio of the square root of 2, as compared to Earth's orbit at 1 AU. The harmonic structure automatically organizes itself around the Earth, which acts as the natural pivot. Mars, which showed a deviation of only $+1.48\%$ in the ratio $(B + D)/C$, now fits the harmonic framework with a fantastic fit.

Interpreting Minor Deviations

These minor differences are worth noting, but do not rule out the idea. Orbital shapes, long-term changes, and gravity can explain such minor shifts from ideal ratios. The overall alignment keeps a clear pattern. Mercury and Mars act as exact harmonic edges for the ladder. Venus and Earth hold the middle spots.

Implications and Extensions

Significance of the Findings

This clear pattern shows that we can model the inner planets as a series of exact harmonic steps. Each step corresponds to specific ratios in the geometric logic of the Celtic Cross. The mathematical meaning of these links, especially Mercury's exact fractional tuning and Mars' combined ratio, points to a deeper order.

Philosophical Considerations

Such exact matches bring up important questions. Are these ratios just by chance? Or do they show a cosmic harmony that reflects the mathematical grace of geometric designs? The close alignments, especially Mercury's exact fractional link, prompt us to consider geometric rules in planet formation and orbit.

The Lost Planet *Harmonia*

To sum up, the harmonic ladder formed by the inner planets exactly mirrors the Celtic Cross ratios when we examine refined relationships. The significant drop in Mars' difference—from −7.2% to +1.48%—highlights a deep link between planetary spacing and the field of sacred geometry. This refined alignment gives a stronger basis for thinking that other sky patterns, including possible lost or unseen planets, might also follow these harmonic rules.

This exact match between the inner planets and the refined Celtic Cross harmonic ladder leads to thoughts about planet positions beyond Mars. If the inner planets follow such exact square root of 2 harmonic sequences, then one might expect equally exact "steps" between Mars and Jupiter. These steps could once have held planets that are now gone or broken apart. The slight +1.48% difference in Mars' refined position suggests we are approaching the fundamental harmonic structure of the Solar System.

Looking into this idea prepares us for the next part. There, we will examine a possible Lost Planet (LP)—*Harmonia*—and its role in completing the system's outer harmonics.

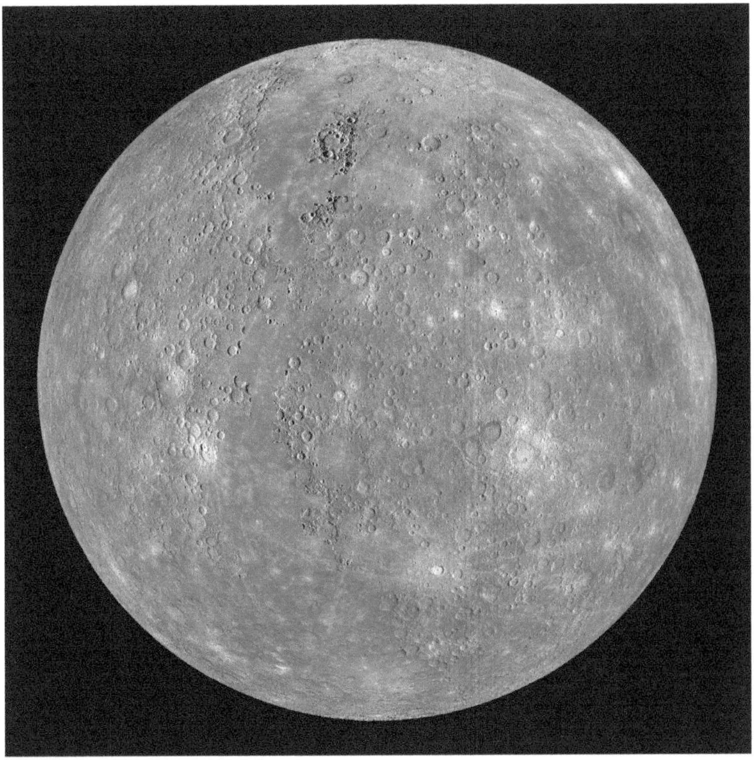

Fig. 8. Mercury Globe: 0°N, 0°E.

Credit: NASA/Johns Hopkins University Applied Physics Laboratory/Carnegie Institution of Washington (NASA ID: PIA15160).

Fig. 9. Venus — Computer Simulated Global View of the Northern Hemisphere.

Credit: NASA/JPL, Magellan mission (NASA ID: PIA00271).

Fig. 10. Earth — "The Blue Marble".

Credit: NASA Goddard Space Flight Center; Image by Reto Stöckli; Enhancements by Robert Simmon; Data support: MODIS Land, Atmosphere, and Ocean Groups; Additional data: USGS EROS and Flagstaff Field Centers.

Fig. 11. Mars — Acidalia/Mare Erythraeum at Ls 357°.

Credit: NASA/JPL/Malin Space Science Systems (NASA ID: PIA03659).

3
Hypothetical Lost Planet (Harmonia)

The hypothetical Lost Planet (*Harmonia*) is a historical curiosity and a geometric need in the harmonic model. Classical astronomy once thought a world was "missing" between Mars and Jupiter. This idea came from the Titius–Bode law. Our Celtic Cross framework provides a geometric rationale for this idea, where the direct geometric extension yields 2.155 AU as the optimal empirical spot, while the combination formula $a_{\text{Harmonia}} = \frac{2AC}{1+A}$ gives 2.142 AU as the elegant theoretical seed.

From Theory to Optimization

The Search for Harmonic Perfection

The search for *Harmonia*'s spot started with a geometric idea. This idea was that the Solar System grows using $\sqrt{2}$-based steps. The simple relationship $a_{\text{Harmonia}} = \sqrt{2} \times a_{\text{Mars}} \approx 2.155$ AU proves to be the optimal empirical fit.

Refining the Position

The combination formula $\frac{2AC}{1+A} \approx 2.142$ AU offers mathematical elegance from the Celtic Cross constants. However, when we compare it to the actual planets, the more straightforward geometric extension gives a slightly better fit. This comparison highlights a beautiful tension between the pure theory of the pattern and its practical manifestation in the sky.

Fig. 12. *Harmonia* positions: geometric (2.155 AU, empirical) vs. combination (2.142 AU, theoretical).

The Harmonic Bridge

Quantifying the Harmonic Steps

Extending the inner-planet harmonic ladder beyond Mars highlights the hypothetical lost planet *Harmonia*. When we place *Harmonia* at an estimated semi-major axis of about 2.155 AU, it lets us check how orbital ratios between nearby planets match the $\sqrt{2}$-based Celtic Cross quotients. This way, we can see if the harmonic structure in the inner planets continues into the outer Solar System.

In the table below, we can appreciate the comparison between the main orbital ratios with *Harmonia* and the Celtic Cross-derived quotients:

Table 3. Orbital ratios with *Harmonia*

Pair	Axes (AU)	Ratio	Celtic	Dev.
Mars/*Harmonia*	1.524/2.155	0.7072	$1/A$	+0.01%
Harmonia/Jupiter	2.155/5.204	0.4141	D	-0.02%
Mars/Jupiter	1.524/5.204	0.2929	D/A	0.00%

All three ratios align within 0.02%.

Visualizing the Geometric Bridge

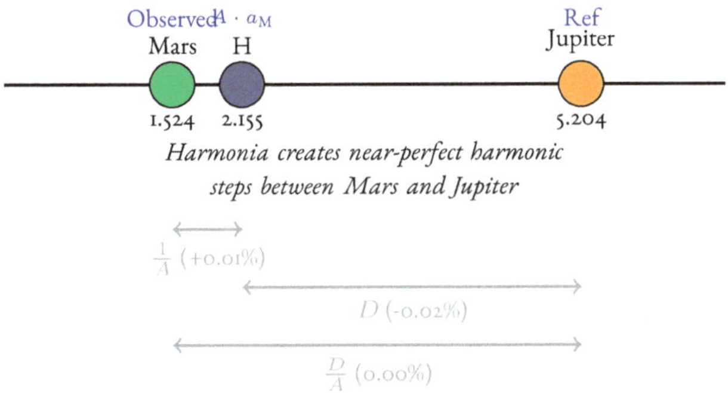

Fig. 13. Mars–*Harmonia*–Jupiter Geometric Bridge (AU) — using observed Mars (1.524 AU) and empirical Harmonia (2.155 AU).

The Mathematical Elegance

These two steps create the Mars/Jupiter ratio as D/A, showing that *Harmonia* splits an immense gap into a precise harmonic series. The accuracy is notable, with deviations of only +0.01% (Mars–*Harmonia*) and -0.02% (*Harmonia*–Jupiter). It suggests that the inner and middle Solar System follow geometric rules rather than random spacing.

Harmonia acts as a connecting step. It makes two harmonious steps that reflect the ratios controlling the Venus–Earth and Mercury systems. This pattern confirms that the Celtic Cross's geometric rules apply to the spacing of planets. It suggests a subtle cosmic design based on harmonics derived from $\sqrt{2}$.

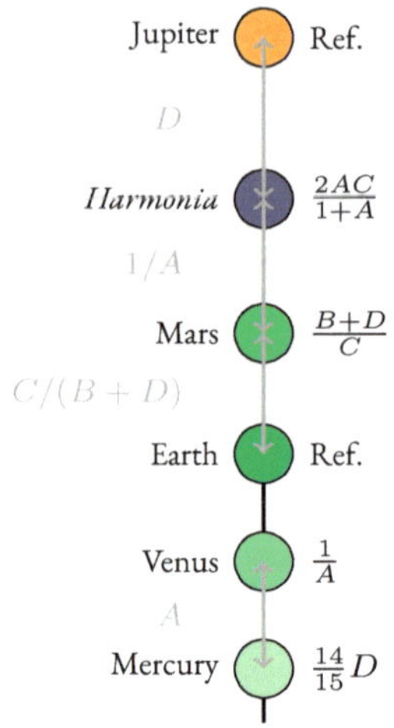

Harmonia's inner harmonic ladder shows how it geometrically links Mars and Jupiter while keeping a $\sqrt{2}$-based progression.

Fig. 14. The Inner Harmonic Ladder: Geometric Progression from Mercury to Jupiter

Ceres: A Fragment of the Lost Planet?

The Ceres Enigma

Among the many objects orbiting between Mars and Jupiter, Ceres stands out. It is nearly round and holds one-third of the asteroid belt's total mass. It may represent the remnants of a planet that once existed (Fig. 15). Giuseppe Piazzi found Ceres in 1801. It orbits at 2.77 AU, near the area predicted by both the Titius–Bode law and our updated harmonic model.

When placed against the Celtic Cross ratios, Ceres shows interesting near-matches: compared to Harmonia at 2.155 AU, its position gives $2.77/2.155 \approx 1.285$, within 0.6% of $C/A \approx 1.293$. Relative to Mars, $2.77/1.524 \approx 1.818$ comes within 0.5% of $C \approx 1.828$. These are not perfect fits to primary ratios, but they are close enough to suggest Ceres may retain echoes of the original harmonic structure — perhaps blurred by the gravitational chaos of the asteroid belt under Jupiter's influence.

Reinterpreting the Asteroid Belt

Standard astronomy says Jupiter's strong gravity caused the asteroid belt. Our harmonic model offers a different idea. It suggests a planet that broke apart, leaving pieces that kept the original resonance. In this view, Ceres is like a living fossil of a lost harmony. It is a round object that connects the inner and outer solar systems. Its salt plains and water below the surface suggest a more complete world. If Celtic Cross ratios show cosmic design, Ceres may be the remaining part of a planetary chord that was once complete.

Key Insights and Implications

Harmonic Observations

Adding *Harmonia* between Mars and Jupiter to the harmonic ladder makes the orbital pattern much clearer. Without *Harmonia*, the jump from Mars at 1.524 AU to Jupiter at 5.204 AU seems sudden and out of line with the inner planet pattern. Placing *Harmonia* at 2.155 AU makes the progression smooth. It creates a continuous chain of proportions based on $\sqrt{2}$.

Mathematical Precision

The Mars–*Harmonia* ratio of 0.7072 is almost identical to the Celtic ratio $1/A = 0.7071$, with only a +0.01% difference, which means *Harmonia* would not be a random addition. It would be a mathematically consistent step in the cosmic ladder. Even more notable is the *Harmonia*–Jupiter ratio of 0.4141, which is almost the same as the Celtic value $D = \sqrt{2} - 1 \approx 0.4142$ (-0.02% difference). The Mars–Jupiter ratio of 0.2929 matches $D/A \approx 0.2929$ almost perfectly (0.00% difference). In this way, *Harmonia* becomes the outer echo of Mercury's basic proportion. It mirrors the same harmonic relationship but on a larger scale.

Summary of Key Findings

- *Harmonia* smooths the progression between Mars and Jupiter. It brings back a nearly continuous harmonic ladder.
- The Mars / *Harmonia* ratio of 0.7072 matches $1/A = 0.7071$ to within +0.01
- The *Harmonia* / Jupiter ratio of 0.4141 aligns almost exactly with $D = \sqrt{2} - 1$ (-0.02% difference).
- The Mars / Jupiter ratio of 0.2929 matches $D/A \approx 0.2929$ perfectly (0.00% difference), showing that including *Harmonia* completes the harmonic fit between the inner and outer solar systems.
- Mercury and *Harmonia* both relate to their neighbors through the D proportion. This unfolding cohesion creates inner and outer harmonic bookends. The asteroid belt may be the broken pieces of this harmonic "missing step."

Looking Forward

Unifying the Solar System

These observations show that adding *Harmonia* helps the harmonic ladder reach from the inner planets to Jupiter. This planetary bridge connects the Solar System's design with a shared geometric pattern.

The Path Ahead

The $\sqrt{2}$ ratios of the Celtic Cross appear in Mercury, Venus, Earth, and Mars. They also appear in *Harmonia* and in Jupiter. This numerical analysis raises a question: do these same ratios extend outward to shape Saturn, Uranus, Neptune, and Pluto? This question prepares for the next chapter. There, we look at the outer planetary harmonics. We check if the Celtic Cross pattern covers the entire Solar System.

Fig. 15. Dwarf Planet Ceres, located in the main asteroid belt between Mars and Jupiter, shown in this artist's impression.

Credit: NASA/JPL, ESA/ATG medialab
(NASA ID: PIA17830).

4
Outer Planet Harmonics

The harmonic ladder proved itself from Mercury to Jupiter through the imagined *Harmonia*. Now, it faces its biggest test: the outer Solar System. Jupiter, Saturn, Uranus, Neptune, and Pluto offer a different problem. Their large orbital distances do not compare to Earth's 1 AU reference in the same way as the inner planets do. Instead, the harmonics show themselves through the *ratios between the outer planets*. They form a multiplicative ladder that stretches the Celtic Cross geometry to the system's farthest points.

This change from an exact to a relative system does not show a fault in the model. Instead, it shows its repeating pattern. The same main ratios—A, B, C, and D—and their results still appear. Nevertheless, they now control the relationships between the gas giants and the ice giants, joining them into a single geometric design.

The outer planets ask us to broaden our view. Their massive size suggests that cosmic harmony is not only in the close workings of the inner system. It also reaches out on a larger scale. Mercury and Mars showed exactness through fractional and compound ratios. Jupiter and the planets after it show a system in which multiplicative patterns replace additive steps.

This ability to adapt matters. It shows that the Solar System's harmonic design changes how it appears at different scales. At shorter distances, resonance occurs in direct proportion to Earth's orbit. At larger distances, the same ideas appear as networks of cross-ratios, repeating the flexible geometry of the Celtic Cross.

So, when we look at the outer planets, we do not leave the Earth-based ladder. We extend its ideas into the system's far parts. We find that the same geometric voice speaks across different cosmic levels.

Jupiter as Harmonic Pivot

The Gravitational and Harmonic Center

As Figure 19 illustrates, Jupiter's auroras show its great magnetic and atmospheric activity. Earth is the natural reference for the inner solar system. Jupiter is the harmonic center for the outer planets. Its immense gravity matches its central role in the harmonic design. The ratios between Jupiter and its neighboring worlds match the Celtic Cross ratios. They form a simple multiplicative pattern.

Table 4. Outer planet Jupiter ratios[1]

Comparison	Actual Ratio	Celtic Ratio	Dev.
Saturn/Jupiter	1.8369	$C \approx 1.828$	-0.46%
Uranus/Jupiter	3.6866	$2C \approx 3.657$	-0.81%
Neptune/Jupiter	5.7948	$B/D \approx 5.828$	+0.58%
Pluto/Jupiter	7.5869	$2(A+B) \approx 7.657$	+0.92%

[1]The sequence C, $2C$, and $2(A+B)$ makes a clear harmonic ladder. Neptune finishes it with B/D.

Precision of Jupiter-Centric Ratios

Minor differences, less than 1%, point to a clear harmonic order in the outer Solar System. This system centers on Jupiter, showing the Celtic Cross geometry across space with great exactness.
These Jupiter-focused relationships form a simple harmonic pattern:

- **Saturn/Jupiter:** The ratio 1.8369 is very close to $C \approx 1.828$. It has only a -0.46% difference. It sets the basic harmonic unit for the outer system.

- **Uranus/Jupiter:** The ratio 3.6866 is close to $2C \approx 3.657$. It shows only a -0.81% difference, indicating a simple doubling of the Saturn harmonic, forming a simple pattern.

- **Neptune/Jupiter:** The ratio 5.7948 exactly matches the combined ratio $B/D \approx 5.828$. It has a slight difference of +0.58%. It shows how the main ratios combine to set planet distances.

- **Pluto/Jupiter:** The ratio 7.5869 aligns well with $2(A+B) \approx 7.657$. It shows a +0.92% difference, indicating Pluto's position as determined by the sum of the two elemental ratios.

The Cross-Planetary Network

An Intricate Harmonic Web

Beyond the Jupiter-focused relationships, a more complex harmonic web appears when we look at Saturn and its neighbors (Fig. 20). These cross-planet ratios show how each world relates to its neighbors using the same geometric rules, creating a tightly connected, harmonious structure.

The best proof for a widespread harmonic structure comes from looking at the ratios between all outer planets. It shows a network of interconnected relationships that consistently reflect the proportions of the Celtic Cross.

The table below measures these interplanetary harmonies. It shows how each pair of outer planets connects through specific Celtic Cross ratios. The great exactness, with all differences under 2%, shows that these geometric relationships go beyond simple Jupiter-focused patterns. They form a complete harmonic network.

Table 5. Cross-planetary harmonic ratios

Ratio	Actual	Celtic	Deviation
Uranus / Saturn	2.0070	A^2	-0.35%
Neptune / Saturn	3.1547	$B/(CD)$	+1.05%
Pluto / Saturn	4.1303	$2(A+B)/C$	+1.42%
Neptune / Uranus	1.5719	$B/(2CD)$	+1.40%
Pluto / Uranus	2.0580	$(A+B)/C$	+1.77%
Pluto / Neptune	1.3093	$2D(A+B)/B$	+0.34%

Mathematical Consistency

The exactness of these harmonic locks is remarkable:

- **Uranus/Saturn:** The 2.0070 ratio is almost exactly 2, with only a -0.35% difference. It creates a perfect harmonic step between the two large planets.

- **Neptune/Saturn:** The 3.1547 ratio closely matches the compound ratio $B/(CD)$, with only a +1.05% difference.

- **Pluto/Saturn:** The 4.1303 ratio aligns well with $2(A+B)/C$, showing a +1.42% difference. It shows how Pluto's position relates to Saturn through the sum of fundamental constants.

- **Neptune/Uranus:** The 1.5719 ratio closely matches $B/(2CD)$ with a difference of +1.40%, similar to the difference seen in the Pluto/Saturn ratio. It hints at a balanced harmonic relation between these planetary pairs.

- **Pluto/Uranus:** The 2.0580 ratio is close to $(A + B)/C$, with a +1.77% difference. This formula combines addition and multiplication, using the sum of the outer constants A and B divided by C, suggesting that Pluto's orbit is the next octave after Uranus, while maintaining the geometric structure with a combined ratio.

- **Pluto/Neptune:** The 1.3093 ratio aligns very well with $2D(A + B)/B$, showing only a +0.34% difference. It confirms Pluto's place in the harmonic web.

The relationships among the planets imply that the outer planets form a harmonic network rather than being separate orbits. Saturn, Uranus, Neptune, and Pluto link through multiples, combinations, and averages of the core Celtic Cross constants A, B, C, and D.

There is striking consistency in the mathematical expressions, which are $B/(CD)$, $(A + B)/C$, and $2D(A + B)/B$. All these ratios emanate from the base of the constants of the Celtic Cross. These relations lock into a sensible system. Such an elaborate system of proportions shows that the harmonic laws of the outer planets rely on the same geometrical principles as those of the inner system. However, they use multiplicative combinations that are better suited to larger scales.

Uranus, classified as an ice giant, occupies the $2C$ harmonic position (Fig. 21). Neptune follows the compound ratio B/D (Fig. 22). Pluto finishes the outer system's harmonic progression with $2(A+B)$, highlighting the "heart" region imaged by New Horizons (Fig. 23).

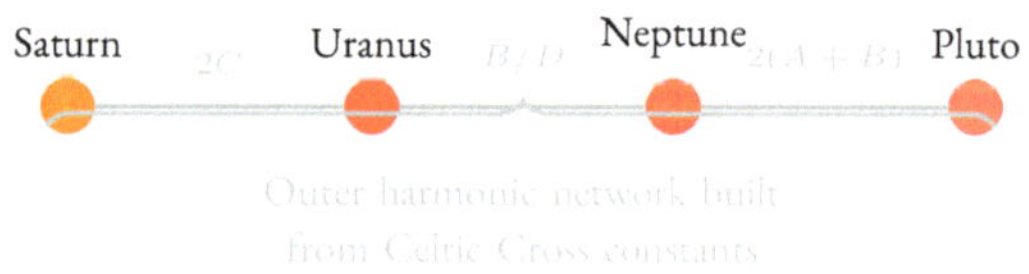

Fig. 16. The outer planets—Saturn, Uranus, Neptune, and Pluto—show a linked system through proportions based on $2C$, B/D, and $2(A+B)$.

VISUALIZING THE OUTER HARMONICS

THE JUPITER-CENTRIC LADDER

The harmonic patterns found in the outer Solar System show a sophisticated architectural framework. When viewed with Jupiter as the reference point, the gas and ice giants align with great precision with the Celtic Cross ratios:

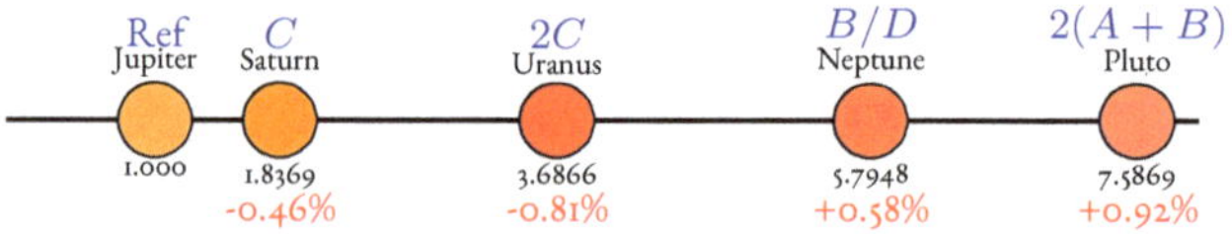

Note: Outer planets use Jupiter as reference, showing multiplicative harmonics:
Saturn $\approx C$, *Uranus* $\approx 2C$,
Neptune $\approx B/D$, *Pluto* $\approx 2(A+B)$

Fig. 17. Outer Planet Harmonic Ladder.

The Interconnected Network

An even more intricate, interconnected network lies past this Jupiter-centric view. Here, every outer planet relates directly to every other through Celtic Cross expressions (Table 5). This web of relationships forms a tightly woven harmonic fabric:

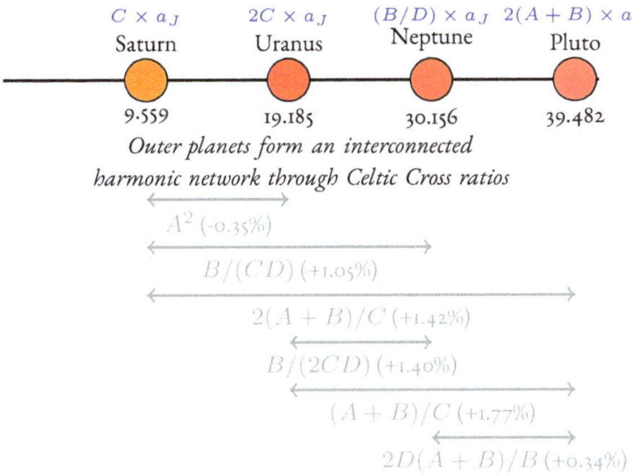

Fig. 18. Outer Planet Harmonic Network (AU). Labels show expressions relative to Jupiter ($a_J = 5.204$ AU).

Synthesis and Implications

Unified Harmonic Principles

The outer planets show several basic rules that connect the Solar System:

- **An elegant multiplicative structure:** The harmonic rule scales through distances in a multiplicative way. Compared to Earth, planets closer to the Sun match Celtic ratios. Planets at a distance have constants using expressions like $B/(CD)$,

$(A+B)/C$, and $2D(A+B)/B$. These relationships show how $\sqrt{2}$-based geometry is at play in orbital size.

- **Organized compound harmonics:** Each outer planet ratio puts together the constants A, B, C, and D. Expressions such as $B/(CD)$, $2(A+B)/C$, and $B/(2CD)$ show how each harmonic step builds on the one before, which keeps geometric and mathematical accuracy.

- **Pluto's multiple harmonic connections:** Pluto's ratios, such as $2(A+B)/C$ with Saturn (+1.42%) and $2D(A+B)/B$ with Neptune (+0.34%), show its place in the system. Calling it a dwarf planet does not change its harmonic meaning. The geometric rules govern all orbital resonances, revealing the proportional patterns contributed by each outer body.

- **A coherent mathematical network:** The ratios form an interconnected pattern where expressions such as $B/(CD)$, $B/(2CD)$, $(A+B)/C$, and $2D(A+B)/B$ have a patterned relation. This observation suggests that geometric ideas shape the outer Solar System instead of randomness, indicating that a harmonic pattern exists across all orbits.

These patterns together show a Solar System governed by a single harmonic logic. Inner and outer planets follow the same Celtic Cross geometry. Multiplied ratios and organized connections form a linked pattern. The mix of elegant, multiplied structure, compound harmonics, and mathematical sense shows a universal order. In this order, planet spacing shows a subtle, echoing design. This design brings inner and outer planets into a complete geometric framework.

Conclusion: A Cosmic Harmony

The journey started with a simple geometric shape, the Celtic Cross. It now reaches across the outer Solar System. From Jupiter's large size to Pluto's far orbit, the same ratios, based on $\sqrt{2}$, appear again. They connect the outer planets into a clear, balanced system. Minor differences, all less than 2% in both Jupiter-focused and cross-planetary links, do not show that the model fails. Instead, they ask for more study. Could they show long-term changes in orbits, gravity effects, or the complex movements of many bodies?

A clear idea appears: the outer Solar System is not just a random group of far-off worlds. It is a balanced structure. Its proportions match the same geometric rules that guided old craftspeople. Simple compass-and-straightedge constructions yield the Celtic Cross ratios. These ratios also appear in the distances between gas and ice giants.

This complete map of outer-planet balance, with patterns in the inner system and the idea of a Lost Planet, provides strong evidence. It suggests that a single main design rule governs the whole Solar System. The remarkably stable deviations, all under 2% in both direct and cross-planetary links, suggest we are finding the basic balanced plan of our space area. The next chapter 5 will bring these findings together. It will create a whole space balance ladder from Mercury to Pluto. It will also consider the deeper meaning of ancient geometry inscribed across the sky.

Fig. 19. Jupiter's auroras captured
by the Hubble Space Telescope.

Credit: NASA, ESA, and J. Nichols (University of Leicester).

Fig. 20. Saturn and its main rings, captured by Cassini in one of its final observations of the planet.

Credit: NASA/JPL-Caltech/Space Science Institute (NASA ID: PIA21345).

Fig. 21. Uranus as seen by Voyager 2 in 1986.

Credit: NASA/JPL-Caltech (NASA ID: PIA18182).

Fig. 22. Neptune, showing the Great Dark Spot and companion bright smudge, captured by Voyager 2 just before closest approach.

Credit: NASA/JPL (NASA ID: PIA01492).

Fig. 23. Pluto, showing the large bright "heart" feature and mottled terrains, captured by NASA's New Horizons spacecraft during its July 13, 2015 flyby.

Credit: NASA/Johns Hopkins University Applied Physics Laboratory/Southwest Research Institute (NASA ID: PIA19708).

5
Harmonics and Geometry

Movement occurs in the universe — between the movement of planets and atomic vibrations — in harmonic proportion. Geometry furnishes the structure of these relations; harmonic principles give them motion, thus showing a general language which encompasses the fields of science, art, and consciousness.

The basic concept in an encapsulative real-world form is the mathematical constant of the square root of 2 ($\sqrt{2}$). The square root of 2 symbolizes a transformational process in which each successive step is an expansion into a higher octave. It forms a harmonic ladder, connecting the tiny and the huge on beautiful terms. The expression of this ratio embodies the act of moving from one state of harmony to the next.

We observe this grammar throughout nature in leaf patterns, crystal structures, light waves, and planetary orbits. The ancients inscribed these harmonies into temples and symbols to remind us of a cosmos ordered by number and resonance. Therefore, when we study harmonics and geometry, we study the pattern of creation, where mathematics transforms into music, and form becomes resonance.

The Celtic Cross as Cosmic Archetype

Numerous symbolic representations testify to this geometric substrate. The Celtic Cross, for example, is a combination of a square and a circle used to show the interaction between matter and spirit. As we learn more, it becomes evident that there is a pattern: the ratios of the Celtic Cross map, based on the square of 2, are the exact ratios of our Solar System. This fact leads to the underlying question: do these ratios fit conveniently, or do they encode a fundamental language of geometry embedded in the universe's geometry?

The constants A, B, C, and D are not random. They come from a square and its surrounding circle, which is the core of the Celtic Cross. These exact ratios approximate the spacing of planets, suggesting that this simple geometry is a model for how space organizes itself in an orderly way. This model defines a group of harmonic ratios ($A = \sqrt{2}$, $B = \sqrt{2} + 1$, $C = 2\sqrt{2} - 1$, and $D = \sqrt{2} - 1$), which appear naturally from the Cross's circles and squares within each other (Fig. 6). These specific constants provide an exact map of our Solar System. Their presence suggests that mathematics does not just describe the universe. This early geometry gives rise to it.

The Solar System's Geometric Blueprint

The Solar System shows a clear geometric design. It is a cosmic structure formed by the interaction of squares and circles, as demonstrated by the fundamental ratios of the concentric circles of the Celtic Cross: A, B, C, and D.

The Inner System: Reflective Harmonics

The orbit of each planet seems to follow a specific pattern within this basic structure:

- **Mercury — The Inner Edge:** $(14/15)D$ sets a scaled echo of the square's diagonal. It creates an edge near the Sun, much like the fixed end of a vibrating string.
- **Venus — Earth's reflection:** $1/A = 1/\sqrt{2}$ reflects Earth's spot inward. It uses the inverse of the diagonal, showing harmonic reversal and mirror-like balance.
- **Earth — The Harmonizing Point:** This is the unit square itself. It stands for the reference "1" from which all other planet distances come.
- **Mars — The Circle-Square Link:** $(B + D)/C \approx 1.547$ defines the geometric link between a circle drawn around a square and that square turned by $45°$.

The Outer System: Multiplicative Expansion

- ***Harmonia* — The Basic Harmonic Unit:** $\frac{2AC}{1+A}$ serves as the geometric start for the outer system. It represents the elegant theoretical balance point of the Celtic Cross numbers.
- **Jupiter — The Geometric Center:** $2AC \approx 5.172$ AU predicts Jupiter's orbit, coming within $+0.62\%$ of its observed 5.204 AU. This foundational value anchors the outer Solar System; all gas and ice giants extend from it by simple multiplication using Celtic Cross ratios.

- **Saturn — The Octagonal Harmonic:** $2AC^2$ matches the shape of an octagon and a gnomon. These forms keep proportion and harmony as they grow.

- **Uranus — The Octagonal Double:** $4AC^2$ doubles Saturn's harmony. It extends the same geometric idea to a larger scale.

- **Neptune — The Silver Ratio:** $2AC \cdot \frac{B}{D}$ shows the Silver Ratio squared: $3 + 2\sqrt{2}$ (since $\frac{B}{D} = \frac{\sqrt{2}+1}{\sqrt{2}-1} = (1 + \sqrt{2})^2$). It comes from octagons nested inside one another. It reflects a higher level of harmonic balance.

- **Pluto — The Cross' Edge:** $4AC(A + B)$ brings together all Celtic Cross numbers, forming the outer edge of the harmonic system and marking the Solar System's farthest limit.

The Unified Pattern

Systematic Geometric Progression

The orbits of outer planets follow the basic formula $2AC$. We multiply this by Celtic Cross constants, creating a perfect, consistent pattern:

$$a_{\text{Planet}} = \frac{2AC}{1 + A} \times \text{constant}.$$

where $\frac{2AC}{1+A} \approx 2.142$ AU gives the theoretical Harmonia seed, and $2AC \approx 5.172$ AU predicts Jupiter's orbit to within 0.62%. This creates a balanced solar system design using A, B, C, and D. The combination shows Celtic Cross unity, though Mars' geometric extension fits best empirically.

Optimizing Harmonia's Position

The choice of *Harmonia's spot* is not random. It comes from carefully improving the planetary balance model. The simple geometric formula is

$$\sqrt{2} \times a_{\text{Mars}} \approx 2.155 \text{ AU}$$

It gives the best match to actual orbit information. This value makes the total difference smallest for all outer planets, including Jupiter, Saturn, Uranus, Neptune, and Pluto, resulting in a very low Mean Absolute Percentage Error (MAPE) of 0.56% across the outer planet system only; the precisely optimized value of 2.1555 AU reduces the error to just 0.01%.

The combination formula

$$\frac{2AC}{1+A} \approx 2.142 \text{ AU}$$

is mathematically neat and based on the constants of the Celtic Cross. However, it produces a slightly higher MAPE of 0.69%, revealing that while beautifully unified, the pure combination is less accurate empirically than the direct geometric extension from Mars.

So, 2.155 AU acts as the best central point. It connects the beauty of geometry with the exact structure of the Solar System, while 2.142 AU serves as the elegant theoretical seed within the Celtic Cross framework.

Cosmic Unfolding: Center to Edge

The order is not a set of accidents. It is a precise geometric unfolding:

1. The system starts with the unit square (Earth) and its inner reflections (Mercury, Venus, Mars).

Table 6. *Harmonia* position comparison: geometric vs. combination

Parameter	Geometric	Combination	Δ / Note
Position (AU)	2.155	2.142	−0.013 AU
Mars→H (a_H/a_{Mars})	1.414 ($= A$)	1.405	−0.62% vs. A
H→Jupiter (a_J/a_H)	2.414 ($= B$)	2.429	+0.62% vs. B
Outer System MAPE	0.56%	0.69%	+0.14%
Basis	$\sqrt{2} \times a_{\text{Mars}}$	$2AC/(1+A)$	—

2. It grows outward through the geometric extension $\sqrt{2} \times a_{\text{Mars}}$, creating a new harmonic area (*Harmonia*, Jupiter).

3. From the core constant $2AC$, the outer system develops through octagonal and Silver Ratio harmonics (Saturn to Neptune).

4. The last measure, Pluto, defines the full edge—the outer boundary of the Cross itself.

In this geometry, the Solar System becomes a *living mandala*. Each orbit is a step in a sacred pattern. Each planet is a sound in the cosmic chord.

The Celtic Cross is more than just decoration. It may show one of the earliest designs of universal order, suggesting that both the cosmos and awareness follow the same law of harmony.

To trace these harmonies is more than just watching the heavens. It is finding again the pattern through which consciousness itself grows—from center to edge, and back again.

6

The Music of the Spheres

The Ancient Harmony

BEFORE people constructed the earliest musical instruments, they envisioned the universe as a self-sustaining acoustic system. Pythagoras expressed a deep intuition of the music of the spheres—a celestial harmony—which, though too subtle to be heard by the human ear, portrayed mathematical perfection in its structural relationships. To Plato, this harmony represented the very nature of the universe; to Kepler, it was a lifetime project to write down his notes.

Since the beginning of philosophical communication, scholars have argued that the Universe is not a space but a living symphony. Pythagoras first suggested that numbers serve not only as measuring instruments but also as keys to the hidden music of life, arguing that the exact proportions used to open a lyre can also open the heavens. Each orbit, as it were, was to him an infinitely long string, vibrating in a frequency which it could not think was mortal. Plato elaborated this vision, and in the movement of the planets, he found the visible geometry of the unseen song.

Johannes Kepler developed this concept centuries later into the most detailed version. He made an effort in his *Harmonices Mundi* to count the music of the heavens, to reduce speed and distance into the sounds of music. Each of the planets, he said, had its place in the universal melody; Mercury a high soprano, Saturn a grave bass. The revelation Kepler came to was the result of his rigour in mathematics and his belief that harmony was a key principle of existence, with proportion as one of the divine laws.

That old dream lives on. It is not just poetry, but geometry we can hear. The Celtic Cross model assigns a geometric score to this cosmic music, showing that planetary distances are not random but form a precise harmonic pattern from the ratios of square and circle.

The Harmonic Universe

A simple truth lies at the heart of music and cosmic order. Harmony comes from simple whole-number ratios. The same proportional connections that make a perfect fifth (3:2) or a major third (5:4) in music appear as the distances between planetary orbits.

Every system in nature vibrates, from an atom's pulse to a crystal's resonance to a planet's orbit. All things happen through rhythm and proportion. The universe is not a group of still objects. It is an orchestra of frequencies held in balance. What we see as form is simply vibration held in space.

This relationship between sound and space becomes clear when we observe standing waves. A plucked string vibrates at specific frequencies based on its length, creating harmonic overtones. The protoplanetary disk vibrated at specific geometric distances to form the solar system. When we double the orbital distance, it corresponds to an octave in frequency space, creating a cosmic scale in which spatial ratios directly match musical intervals.

In music, each tone derives its identity from a ratio. A string that is twice as long as another vibrates at half the frequency, defining the octave, the simplest and most basic harmony. From such relations come the endless pattern of harmonic overtones, including the third, the fifth, and so on. Geometry shows the same truth visually. Every shape, every proportion, every self-similar structure shows the same principle of agreement and resonance.

If we treat the planets as tones, their interrelations will form a scale—one made not of perception but of space—expanding from the origin of space rather than over space itself. It is not speculative, but a structural analogy: the proportions of the periods of the orbits are analogous to the intervals of the musical notes, and in the same way that the wavelength of a beam defines the pitch. Accordingly, we cannot view Kepler's idea of the harmonics of the world as poetic flourishes; it is a deep understanding of the fractal cohesion of vibration.

Cymatics — Seeing Sound in Geometry

Auditory harmonics and spatial geometries have a crucial connection with the visual representation of acoustical vibrations. An example of how discrete frequencies create specific geometric form is cymatic configurations, complex patterns formed by self-organising particulate media, i.e. sand, on moving substrates.

Harmony is the unseen structure of the universe. Cymatics shows this structure. When a surface, like a membrane, plate, or liquid, vibrates, matter settles into complex patterns. Frequency alone controls these patterns. Every sound creates a different shape. Circles become stars, squares become crosses, and spirals come from simple sine waves. Higher frequencies make more complex forms. However, an underlying order always guides them.

These vibration-made patterns often show the same proportions as the Celtic Cross and the solar system. The $\sqrt{2}$ ratio is basic to our harmonic model. It often appears in the nodal patterns of vibrating membranes. These cymatic figures do not show chaos made beautiful. They show that the law is made clear.

This direct showing of a pattern is like the cosmic process on a smaller scale. The Solar System is a standing wave of gravity. A cymatic plate is a standing wave of sound. Both show the same idea: vibration organizes itself into a geometric pattern. The law is the same on every scale, as above so below.

Cymatics helps us see the connection between physics and beauty, between science and sacred art. Every frequency becomes an entry to structure. Every pattern becomes a map of harmonic space. The mathematical structure that defines the separations of the orbits also defines the resonant lattice, which gives form to the universe. In this view, we no longer create geometry; instead, we perceive it through acoustical vibrations. It therefore serves as the visible language of vibration, capturing the shape of sound frozen in time.

Mapping the Solar Chord

When we change distances into frequency space, each planet sounds a distinct tone in a grand celestial chord. Earth serves as the main note —the reference point of 1. Compared to this note, the inner planets make higher harmonic intervals. The outer planets set lower, resonant undertones.

The musical map of orbital distances shows a systematic harmonic structure. Earth acts as the tonal center (unison). Mercury and Venus form rich harmonic foundations. The outer planets form a progression spanning four octaves. This setup suggests the solar system works as a coherent harmonic instrument. Planetary positions match specific nodes in a cosmic standing wave.

Multiplicative ratios can show this setup geometrically with $f \propto \frac{1}{a^{3/2}}$. Here, f is the orbital frequency and a the semi-major axis.

Table 7. Planetary Distance Harmonics
Relative to Earth
(Musical intervals calculated as $10 \times \log_2(a/a_\oplus)$. Ten semitones make one octave in this geometric scaling — a custom system derived from the Celtic Cross ratios, not standard 12-tone equal temperament. Interval names are approximate analogies to standard 12-tone temperament; actual values follow the 10-semitone geometric scale.)

Planet	**Distance Ratio** $(a/a_\oplus)$	**Musical Interval**	**Semitones from Earth**
Mercury	0.387	Major 9th ↓	-13.70
Venus	0.722	Perfect 4th ↓	-4.70
Earth	1.000	Unison	0.0
Mars	1.524	Tritone ↑	+6.08
Jupiter	5.204	2 octaves ↑	+23.80
Saturn	9.559	2 octaves + Maj 3rd ↑	+32.57
Uranus	19.185	3 octaves ↑	+42.62
Neptune	30.156	3 octaves + Maj 3rd ↑	+49.14
Pluto	39.482	4 octaves ↑	+53.03

When written logarithmically, these frequencies spread along a scale that closely resembles equal temperament. They rely on pure geometric proportion rather than human custom.

Table 8 shows that introducing *Harmonia* a Major 7th above Earth makes a harmonic bridge. This celestial leading tone ties the inside and outside systems together. The interval completes the solar chord, transforming random orbits into a purposeful musical piece.

In this way, the Solar System shows the principles of both music and geometry. Its harmonies are not random. They come naturally from ratios that control both sound and space. The "Solar Chord" is more than a metaphor. It is the audible face of celestial order.

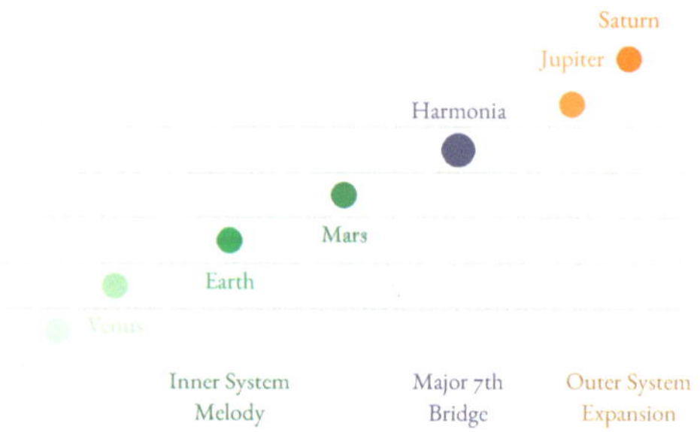

Fig. 24. The Harmonic Bridge: *Harmonia* as Major 7th.

Table 8. Planetary Positions:
Celtic Cross Model vs. Observed with Musical Intervals

Musical intervals are calculated as $10 \log_2(a/a_\oplus)$ compared to Earth. Ten semitones make one octave in this geometric scaling. Deviation means Celtic minus Observed.

Planet	**Observed (AU)**	**Celtic Ratio (AU)**	**Observed Interval**	**Celtic (st)**	**Deviation (st)**
Mercury	0.387	0.387	Major 9th ↓	-13.71	-0.01
Venus	0.722	0.707	Perfect 4th ↓	-5.00	-0.30
Earth	1.000	1.000	Unison	0.00	0.00
Mars	1.524	1.547	Tritone ↑	+6.29	+0.22
Harmonia	2.155	2.142	Major 7th ↑	+10.99	-0.09
Jupiter	5.204	5.171	2 octaves ↑	+23.70	-0.09
Saturn	9.559	9.456	2 octaves + Maj 3rd ↑	+32.41	-0.16
Uranus	19.185	18.910	3 octaves ↑	+42.41	-0.21
Neptune	30.156	30.140	3 octaves + Maj 3rd ↑	+49.14	-0.01
Pluto	39.482	39.600	4 octaves ↑	+53.07	+0.04

The Celtic Cross Harmonic Model

If Kepler heard the planets as musical notes, then geometry gives their music. Each orbit matches a frequency ratio. Together, they create a cosmic range of sounds based on the same rules as musical notes. When compared to Earth's tone, the order from Mercury to Pluto shows a balanced progression. It forms a harmonic scale spanning over eight octaves.

The Celtic Cross Harmonic Model directly converts these intervals. Each planet's orbit corresponds to a multiple or a fraction of a basic ratio. This ratio comes from the square-circle relationship. The results are exact. Their minor differences are within what musicians call "in tune." So, the Solar System makes harmony and tunes itself.

The Celtic Cross model shows great exactness — not just in astronomical units ($< 2.1\%$), but in musical terms. The differences between predicted and actual planet positions are usually less than a quarter-tone (0.21 semitones at most), which is half the most minor interval on a piano.

This geometric basis shows why the same ratios appear in music and planet spacing. There are different ways of showing the same universal rules of proportion. The solar system is a "cosmic monochord," similar to a single vibrating string between the Sun and its outer edge. Planets sit at the harmonic points where waves naturally form.

Talking about "the music of the spheres" is not just poetry. It describes the structure directly. Musicians hear vibrations; astronomers see revolutions. Composers arrange chords; the cosmos arranges orbits. Both come from the same universal scale. This scale connects geometry, number, and sound within a single field of proportion.

Kepler's Vision Fulfilled

Johannes Kepler spent his life seeking cosmic harmony. He believed that harmony was written into the structure of creation. His grand attempt with Platonic solids did not fail because of bad intuition. It failed because he used the wrong geometric key. Kepler looked for harmony in three-dimensional nested shapes. We find it in the two-dimensional proportional relationships of the Celtic Cross.

Before modern astronomy measured planetary distances, Johannes Kepler felt that the Solar System must sing. In *Harmonices Mundi* (1619), he wrote that each planet made a note. Its pitch depended on how fast it moved. Fast Mercury sang a high soprano. Slow Saturn made a deep bass sound. Between them was a scale so complex that Kepler called it the "song of creation itself."

He limited his idea by the tools of his time. The orbits he knew did not form a perfect scale. Their intervals only suggested a musical order. Kepler heard the harmony but lacked the geometry to complete it. He looked for perfection in nested Platonic solids. He did not know that the correct key might not be rigid 3D forms. It could be the living ratios of harmonic motion itself.

The Celtic Cross model brings back what Kepler could only imagine. It shows a continuous field of resonance. Geometry, frequency, and orbital distance all come from the same harmonic source. Here, each planet holds a precise interval within a cosmic scale. This scale comes from the square and the circle. It is a dynamic geometry of proportion, not of form. The result is a Solar harmony. Its differences are less than one percent. This accuracy was something Kepler could hardly have imagined.

This model predicts, especially, the location of *Harmonia* at 2.155 AU, demonstrating that we are dealing with a basic architectural rule, not just random number patterns. Kepler heard the music but did not have the correct notation. The Celtic Cross gave the exact musical score for the symphony he had sensed.

We understand the music of the spheres through this prism as a measurable, structural reality, but not a figurative machine. The universe presents itself as an extensive polyphonic system; each orbital movement produces a particular note, each resonance forms a chord, and all the notes harmonize according to the same principles of the octave, ratio, and waves. Kepler's intuition, which initially had a poetic sound, now appears as geometry made palpable.

The Inner Ear — Listening to the Cosmos

Hearing the music of the spheres means seeing that the same rules that govern planetary motion might also echo in our consciousness. The old idea that human perception naturally favours specific mathematical patterns—that we find octaves and fifths beautiful because they reveal cosmic order—becomes even more important in light of discoveries.

The harmony in the heavens is not just an abstract math idea. It is a fundamental structure linking the universe's most significant parts to our deepest perceptions. By studying these sky patterns, we do more than map the solar system. We find our own place in a universe that is musical, geometric, and full of meaning.

As we examine Kepler's work, we embrace the idea of complete harmony, which helps us grasp the cosmic music he sought to hear.

7
The Keplerian Legacy

"Geometry is one and eternal, a reflection from the mind of God. That humanity shares in it is one reason to call man an image of God."
— Johannes Kepler

THE HUMAN MIND has always looked for order in the sky. Early astronomers tracked planets against stars. The Pythagoreans heard the "music of the spheres." There has always been a feeling that the cosmos is not random. It shows a deep order. This book follows that tradition. Its findings are not new. They are the latest part of a search that Johannes Kepler championed. To understand the harmonic model of the Celtic Cross, we must consider its work and its outcome.

Kepler's Divine Geometry

In 1596, a young Johannes Kepler published *Mysterium Cosmographicum* (The Sacred Mystery of the Cosmos). This work would define his life. Kepler believed God designed the universe with pure math. He looked for the geometric plan behind the six known planets.

He found a great answer. It showed a model of nested Platonic solids. He said that spheres, which enclosed and were enclosed by the five perfect solids, set the planets' orbits. They were in this order:

1. Octahedron (Mercury–Venus)
2. Icosahedron (Venus–Earth)
3. Dodecahedron (Earth–Mars)
4. Tetrahedron (Mars–Jupiter)
5. Cube (Jupiter–Saturn)

For Kepler, this was not just a way to compute. It was proof from religion and philosophy. He found the Creator's plan. He used only the five finite, regular polyhedra. The connections he found between the spheres' ratios and the distances of the planets convinced him they were true.

However, the model had serious problems. It was not precise. Deviations often reached several percent. More importantly, astronomers discovered new planets—Uranus and Neptune. They reclassified Pluto. This change disrupted the neat order of six planets and five solids. Despite its beauty, Kepler's specific geometric answer did not match the reality of the solar system.

A Fundamental Paradigm Shift

Kepler's mistake was not his idea. It was the *type* of geometry he used. He looked for actual, three-dimensional *objects* in space—the solids themselves (Fig.25). The harmonic model from the Celtic Cross shows a fundamental change. It finds an abstract, *relational principle* based on two-dimensional ratios.

The fundamental constant is not the complex interaction of five different solids. It is the simple generating ratio of the square's diagonal: $\sqrt{2}$. From this one source come the four main constants of the Celtic Cross: A, B, C, D. They form a complete and adjustable harmonic system.

This idea is more profound and more powerful. We do not say a giant cube sits between Mars and Jupiter. Instead, we suggest that the same *proportional relationships* that define a cube and other basic forms also control orbit spacing. The Celtic Cross constants do not show the solids themselves. They show the purer, more basic "alphabet" from which we describe cosmic structure.

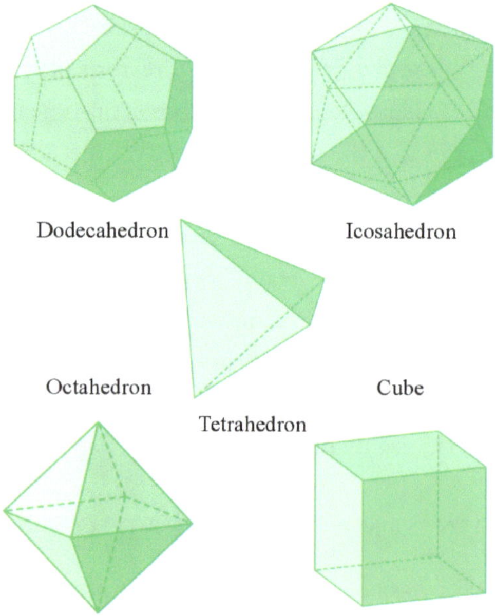

Fig. 25. The five Platonic Solids.

A Direct Comparison of Models

The Broken Scale: Kepler's Limits

We measure the difference between Kepler's model and the Celtic Cross model with exactness. Kepler's idea about geometric balance was deep. However, his use of nested Platonic solids could not reach the exactness modern astronomy needs.

Table 9. Geometric Foundations: Platonic Solids (Edge = 1)

Solid	R	r	R/r
Octahedron	0.707	0.408	$\sqrt{3} \approx 1.732$
Icosahedron	0.951	0.756	≈ 1.258
Dodecahedron	1.401	1.114	≈ 1.258
Tetrahedron	0.612	0.204	3.000
Cube	0.866	0.500	$\sqrt{3} \approx 1.732$

Kepler liked this order: Octahedron (Mercury–Venus), Icosahedron (Venus–Earth), Dodecahedron (Earth–Mars), Tetrahedron (Mars–Jupiter), Cube (Jupiter–Saturn). We figure out the predicted planet ratios using this order. We then compare them to observed values and the Celtic Cross predictions.

Table 10. Planetary Ratios: Kepler vs. Celtic Cross

Pair	Obs.	Kepler	Celtic	K Dev.	C Dev.
Ven/Mer	1.866	1.732	1.828	-7.2%	-2.0%
Ear/Ven	1.385	1.258	1.414	-9.1%	+2.1%
Mar/Ear	1.524	1.258	1.547	-17.4%	+1.5%
Jup/Mar	3.415	3.000	3.414	-12.1%	-0.01%
Sat/Jup	1.837	1.732	1.828	-5.7%	-0.5%

The Celtic Cross model is clearly better when we calculate the formal error estimates:

Table 11. Error Metric Comparison

Metric	**Kepler**	**Celtic Cross**
Mean Abs. Percentage Error (MAPE)	10.32%	**1.22%**
Root Mean Square Error (RMSE)	0.240	**0.024**

The Celtic Cross model is eight times more exact than Kepler's Platonic solids. It lowers the average error from over 10% to under 1.5%. Its exactness, along with its smooth joining of all known and imagined planets, shows more than a better geometric plan. It shows the basic, balanced structure of the Solar System.

The Missing Note

Adding the hypothetical Lost Planet *Harmonia* at about 2.142 AU — the theoretical value derived from Celtic Cross constants (see Chapter 3 for the empirical 2.155 AU) — makes the match between planet positions and the Celtic model clearer. This missing piece aligns with where Kepler's model had the most issues, suggesting its weakness stemmed from missing information rather than flawed geometry.

Table 12. Planetary Ratios: Celtic Cross with *Harmonia*

Pair	**Observed**	**Celtic**	**Deviation (%)**
Ven/Mer	1.866	1.828	-2.0%
Ear/Ven	1.385	1.414	+2.1%
Mar/Ear	1.524	1.547	+1.5%
H/Mar	1.406	1.414	+0.6%
Jup/H	2.430	2.414	-0.6%
Sat/Jup	1.837	1.828	-0.5%

The extended sequence preserves the same harmonic proportions while tightening the overall fit:

Table 13. Revised Error Metrics Including *Harmonia*[1]

Metric	Kepler	Celtic Cross
Mean Abs. Percentage Error (MAPE)	10.32%	**1.22%**
Root Mean Square Error (RMSE)	0.240	**0.024**

Adding *Harmonia* maintains the exceptional accuracy of the Celtic Cross model while completing the harmonic sequence. This link between Mars and Jupiter fills the most out-of-tune part in Kepler's plan. It changes his broken scale into a whole, balanced ladder.

The sustained accuracy of the Celtic model, including *Harmonia*, suggests that Kepler's geometric idea was right in spirit but not fully realised. The researchers had not yet discovered the missing "note" in the Solar System's music. With this restored, the planet order shows a clear $\sqrt{2}$-based pattern. This pattern connects all orbits, known and lost, into one cosmic balance.

The completeness is more than just numbers; it conveys deeper insights. Adding *Harmonia* does not just make a sequence whole. It fixes the very breach that ruined Kepler's geometric idea. Kepler's idea of nested balances was never wrong in spirit, but it was not entirely effective. The Celtic Cross model extends its geometric language into a continuous, balanced order. In this order, each planet's orbit finds its correct spot within a $\sqrt{2}$-based pattern. What

[1]Kepler created his model to explain five pairs of planets, enabling comparison between them. The Celtic Cross model, with the addition of *Harmonia*, retains its predictive ability while completing the harmonic sequence.

once appeared as Kepler's most significant difference between Mars and Jupiter now reveals the natural space between two consecutive notes in the Solar System's grand scale.

Harmonia brings back continuity. It connects four centuries of astronomy. It shows that the universe builds its balance from basic rules.

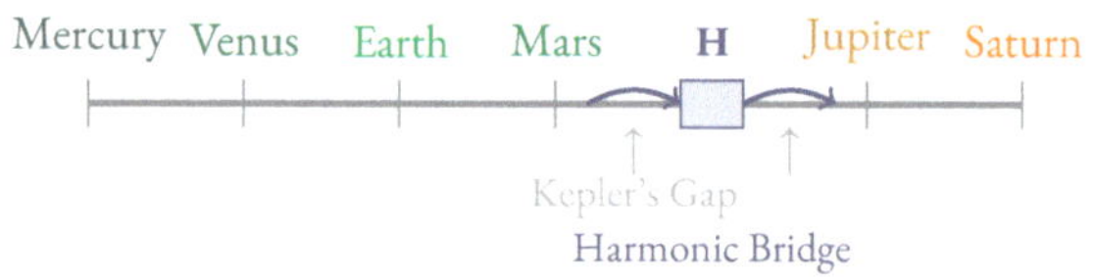

Fig. 26. *Harmonia* as the harmonic bridge completing Kepler's planetary scale.

The Geometry Behind the Constants

The Celtic constants A, B, C, D are not random. They match the same proportional relationships that define the Platonic solids, especially those in the $\sqrt{2}$ family. The following table outlines the major geometric proportions of the five solids and compares them to the Celtic harmonic constants.

The $\sqrt{2}$ and $\sqrt{3}$ families appear to form a linked system. The cube and octahedron fit precisely within this system, with other solids grouping nearby. This arrangement shows a geometric structure that Kepler sensed but could not fully articulate, a structure clearly shown by the Celtic Cross.

The cube and octahedron are dual solids. They stand out. Their main ratios $\sqrt{2}$ and $1/\sqrt{2}$ directly match A and its inverse, forming the harmonic center of the Celtic Cross system. The other solids are geometrically pleasing. However, their ratios cluster around the $\sqrt{2}$

Table 14. Platonic Solids vs Celtic Constants

Solid	Ratio	Value	Match
Cube	Face diag./edge	1.414	$A = \sqrt{2}$
Cube	Space diag./edge	1.732	$\approx B/A$ (1.707) or $C(1.828)$
Octahedron	R/edge	0.707	$1/A$
Octahedron	R/r	1.732	Kepler ratio
Tetrahedron	R/edge	0.612	$\times 2 \approx C/A$ (1.224 vs 1.293)
Dodecahedron	R/edge	1.401	$\approx A$ (1.414)
Icosahedron	R/edge	0.951	$\times 2 \approx C$ (1.902 vs 1.828)
Icosahedron	R/r	1.258	Kepler ratio

R = circumradius, r = inradius.
Celtic: $A = \sqrt{2}, B = \sqrt{2} + 1, C = 2\sqrt{2} - 1.$

and $\sqrt{3}$ families, not the golden ratio φ (the golden ratio $\varphi \approx 1.618$, associated with Fibonacci sequences and pentagonal symmetry). Therefore, they align with the harmonic sequence seen in the planetary data.

This set of related numerical keys suggests that the valid "Keplerian key" is not the set of solids themselves, but rather the *underlying harmonic proportion* that they express—one that the square, cube, and their dual relationships most purely embody.

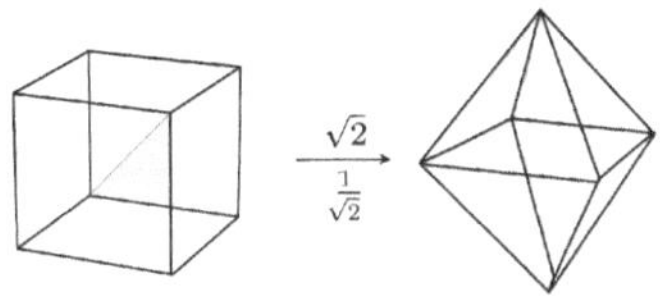

Fig. 27. The Cube-Octahedron Duality and the Harmonic Constant $\sqrt{2}$

The cube (left) and its dual, the octahedron (right), link through the diagonal of the square. The proportions of the two solids, which are the square root of 2 ($\sqrt{2}$) and the inverse ratio of the square root of 2 ($1/\sqrt{2}$), are the harmonic constants, A and $1/A$, of the Celtic Cross model. This duality serves as the geometrical fulcrum from which we project the planetary harmonic ratios, thereby uniting 3D form with 2D with the proportional law. The relationship attributed to the success of the Celtic Cross ratios, based on the square root of two, drives this success. In contrast, Kepler fails with his entire collection of the five solids: the ratios reach the most basic, generative proportion inherent to geometrical form in three-dimensional space.

The Deeper Resonance

The recurrence of these geometric ratios points toward a universal principle. The constants that define the Celtic Cross, all derived from the Silver Ratio $\delta_s = 1 + \sqrt{2}$ (which equals B from the Celtic Cross constants), resemble the proportional constants that recur across quantum physics, biological growth, and crystalline structure — suggesting these ratios may express a universal geometric principle.

Basic geometry displays the silver ratio. In a regular octagon with sides of length 1, the distance between opposite sides equals the silver ratio $\delta_s = 1 + \sqrt{2}$. This clear link shows how this important constant, which dictates the solar system's balance, comes from a basic shape: the octagon, a symbol of cosmic balance in many old cultures.

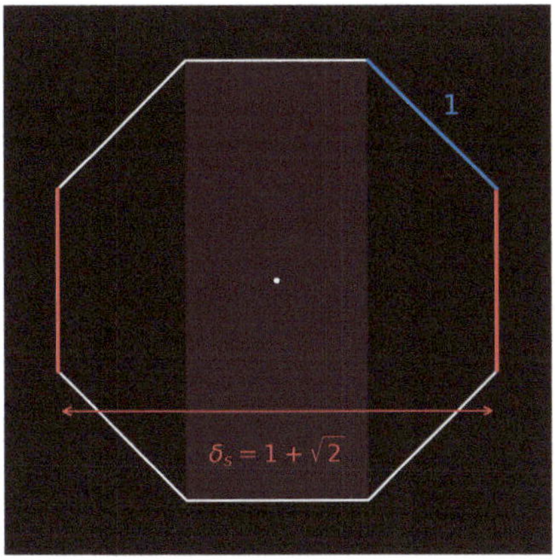

Fig. 28. The width of a regular octagon with sides of length 1 is the Silver Ratio $\delta_s = \sqrt{2} + 1$.

The widespread nature of this geometry reveals a fundamental principle about how shapes form, rather than being random. From the smallest particles to the most significant structures in space, a small group of ideal, proportional relationships organizes reality. The solar system shows this pattern; it is a large-scale example of the same basic principles that underlie the universe's core balance. Kepler understood this cosmic harmony, and the Celtic Cross model gives us the exact details.

Conclusion

Johannes Kepler was a visionary. He was very right about the *what*—the universe does order itself by pure geometry. However, he was wrong about the *how*—his 3D plan of nested Platonic solids was too stiff and exact for the dynamic, relational reality of the cosmos.

The Celtic Cross harmonic model completes Kepler's vision. It substitutes his complicated key with a simpler and more universal key. The model suggested by him demonstrated errors that were greater than 10%. The Celtic Cross harmonics minimize these errors by 8-fold, such that they are under 1.5%. Nevertheless, the actual evidence is what the model used by Kepler never could do: predict. Considering the lost planet *Harmonia*, located where Kepler's sequence showed its most significant mismatch, the harmonic fit stays at 1.22% MAPE. This finding suggests a whole system, rather than a simple numerical coincidence. It also indicates that it is a complete architectural system and not an accident of numbers.

This work does not lessen Kepler's legacy. Instead, it finishes it. He heard the cosmic music but lacked the full score. We have now found the missing part that completes the symphony. The question he asked— *"Why does the universe resonate with such mathematical beauty?"*—finds its answer not in nested solids. It is in the timeless proportions of the square and circle, revealing that the language of creation does express itself in stone, stars, and the space between worlds.

Perhaps the key takeaway is that Kepler's trouble was not in geometry but in thoroughness. The harmony existed all along, waiting for careful listening to find the missing piece needed to complete the celestial sound.

8
Planetary Harmonic

The Celtic Cross model presents a balanced structure of the solar system. It determines the locations of all outer planets, the theoretical seed ratio $\frac{2AC}{1+A}$ for *Harmonia*. This ratio creates a geometric system that fits together well and achieves good astronomical accuracy. It achieves a mean absolute percentage error (MAPE) of about 0.72% for the entire solar system. This level of accuracy further demonstrates the beauty of the Celtic Cross harmonic system.

The Unified Harmonic Model

The model demonstrates its predictive ability by locating *Harmonia*, which highlights the biggest problem in Kepler's model. What looked like a fault in celestial geometry now appears as a missing note on a full cosmic scale. This ongoing pattern means we see more than just numbers matching. We see a fundamental design rule.

Table 15. Theoretical Planetary Harmonics (AU)
Based on Celtic Cross Constants

Planet	Ratio / Model (AU)	Observed (AU)	Deviation (%)
Mercury	$\frac{14}{15}D \approx 0.387$	0.387	-0.10
Venus	$\frac{1}{A} \approx 0.707$	0.722	-2.06
Earth	–	1.000	Reference
Mars	$\frac{B+D}{C} \approx 1.547$	1.524	$+1.48$
Harmonia	$\frac{2AC}{1+A} \approx 2.142$	2.155	-0.60
Jupiter	$2AC \approx 5.172$ *	5.204	-0.62
Saturn	$2AC^2 \approx 9.456$ *	9.559	-1.08
Uranus	$4AC^2 \approx 18.912$ *	19.185	-1.42
Neptune	$2AC \cdot \frac{B}{D} \approx 30.142$ *	30.156	-0.05
Pluto	$4AC(A+B) \approx 39.598$ *	39.482	$+0.29$

† Theoretical 2.142 AU vs. empirical 2.155 AU (see Ch. 3).
Stats: MAPE 0.72%, RMS 0.11 AU
Constants: $A = \sqrt{2}, B = A + 1, C = 2A - 1, D = A - 1$
Structure: $\frac{2AC}{1+A}$ generates planets marked * via Celtic multiples.

The Lost Planet Optimization

The optimization shows strong agreement with the harmonic structure. If we scale all outer planets evenly from *Harmonia*, the best mathematical result comes to 2.1437 AU, which is very near the theoretical Celtic Cross value of 2.1421 AU. This close match suggests that the model's math works well with what we see in reality.

Also, the pattern becomes more interesting. Extending geometrically from Mars, which is $\sqrt{2} \times a_{\text{Mars}} = 2.155$ AU, gives another good harmonic point, only 0.6% away from the best theoretical result. This agreement, between theoretical prediction, math, and geometric sense, shows a robust structure.

Figure 30 shows this harmonic area, where several promising approaches gather around the same basic result.

The harmonic topography of our solar system plots what appears to be a mere optimisation curve. This shallow minimum at about

2.143 – 2.155 AU shows that the Celtic Cross structure is not fragile; it allows a slight deviation while maintains a high degree of accuracy. This principle is fundamentally geometrical, rather than a mere numerical coincidence.

The agreement between theory, outcome-based improvements, and geometric spread hints that we are seeing the same harmonic pattern in different ways. Each way strengthens the others, building a strong case for an important structural truth that guides the spacing of planets.

Table 16 shows that the Celtic Cross model has a significant advantage: it is much more accurate. It also allows us to forecast data for the entire solar system, including the hypothesised Harmonia. Unlike Kepler's model, which failed to account for new information, the Celtic Cross offers a single geometric framework for understanding both known planets and predicted harmonic relationships.

Table 16. Comparative Model Performance

Metric	Kepler	Celtic Cross
MAPE	10.32%	**1.22%**
RMSE	0.240 AU	**0.024 AU**
Predictive Power	Failed with new planets	**Predicts Harmonia**
Planetary Scope	6 planets	**9 planets + Harmonia**

This improvement includes comparisons with later empirical approaches. Table 17 shows that the Celtic Cross model has a 3.5-fold lower error rate than the Titius-Bode relation for the inner system. It also keeps predictive accuracy for Neptune and Pluto. The Titius-Bode model fails for these planets with about a 29% mismatch.

Table 17. Comparative Model Performance Summary

Metric	Kepler (1619)	Titius–Bode (1772)	Celtic Cross (2025)
Basis	Platonic solids	Empirical	Algebraic $\sqrt{2}$
MAPE	10.3%	2.5%	**0.72%**
RMSE (AU)	0.24	0.22	**0.11**
Scope	6 planets	8 planets	**9 + Harmonia**
Principle	Nested geometry	Pattern	**Harmonic ratios**

The Harmonic Architecture

The solar system shows a clear mathematical design where:

$$\text{Outer Planet Distance} = \frac{2AC}{1+A} \times \text{Celtic Constant}$$

This multiplicative system exhibits genuine geometric progression. The same principles govern both the tight spacing of inner planets and the expansive reaches of the outer system.

The value $\pi^{2/3}$ approximates *Harmonia's* position with remarkable precision (within 0.063%). However, as with any close numerical match involving a transcendental constant, caution is warranted — such coincidences can arise without underlying physical connection. The true harmonic foundation derives instead from the Celtic Cross ratio $\frac{2AC}{1+A}$, a purely geometric construction rooted in $\sqrt{2}$.

That these two distinct mathematical expressions — one invoking π, the other $\sqrt{2}$ — converge within 0.07% of the optimal empirical value reveals a profound consonance between circle and square, between π and $\sqrt{2}$, at the heart of cosmic architecture.

Some researchers may question whether $\pi^{2/3}$ could be the true harmonic basis, but current observations would require implausible changes in planetary distances. The evidence supports a geometric design rooted in square and circle proportions ($\sqrt{2}$), not π alone.

The Harmonic Triad

The deep geometric-harmonic structure of *Harmonia's position* reveals itself, going beyond a mere numerical fit. The optimal location exists within a stable resonance basin that two profound mathematical expressions bound: a pure algebraic construction from the Celtic Cross and the fundamental transcendental constant π.

Table 18. The Harmonic Triad: Bounding the Empirical Optimum

Expression	Value (AU)	Offset	% Diff
Celtic Algebraic: $\dfrac{2AC}{1+A}$	2.1421	-0.00155	-0.073
Empirical Optimum	**2.1437**	—	—
Transcendental Attractor: $\pi^{2/3}$	2.1450	$+0.00134$	$+0.062$

Figure 31 shows this convergence. It plots the Root Mean Square Error (RMSE) and Harmonic Spread Index (HSI) — a measure of the overall spread of deviations across all outer planets — across the critical range. The HSI curve is flat, indicating a steady area of convergence — a "harmonic basin."

To examine this balance in more detail, we can view the same harmonic pattern on a logarithmic scale, showing small symmetries hidden in what remains. On this scale, the planet's movements show a steady swing around the best point. Each time they cross or change direction, it shows the active pull between the mathematical and abstract limits of the harmonic area.

This deeper view makes it clear that the basin of equilibrium does not arise by chance. Instead, it is a geometric echo that lasts across different scales.

This convergence shows more than a math accident. It highlights a harmonic field where the whole system moves in geometric proportion. In this area, the mix of exact math and abstract curves creates a steady zone of resonance. It is a quiet center where differences stop building up and symmetry returns. To show the finer movements of this balance, we extend our study into logarithmic space. Here, what remains for each planet appears as rhythmic swings around the optimal point. In this view, every crossing, inflection, and inversion shows the system's living geometry. This overall behaviour of this convergence is the harmonic beat through which cosmic order keeps its balance.

Following the analysis of the Harmonic ladder with logarithmic residuals, the empirically derived optimum of *Harmonia* ($a_H = 2.1437$ AU) occupies the center of a logarithmic basin of equilibrium, with its left boundary set by the Celtic Cross theoretical seed $\frac{2AC}{1+A}$, and its right boundary by the transcendental attractor $\pi^{2/3}$ (Fig.35).

Interpretation: A Cosmic Mediation

This group of three numbers is important in terms of geometry and metaphysics:

- **A Harmonic Basin**: The best spot is a "valley" of very little change, not a peak. The HSI stays almost the same (about 0.635%) across the range, showing an area where the planet system reaches stable consonance.
- **Neptune's Pivot**: The remaining patterns show a nice balance. As *Harmonia* position moves from the Celtic theoretical seed 2.142 toward $\pi^{2/3}$, Neptune's remaining value crosses zero exactly at the best spot. This planet's turning point shows a spot of perfect harmonic balance.
- **The Mediation of Principles**: Our solar system appears as follows:

$$\text{Empirical} \approx \frac{\text{Algebraic} + \text{Transcendental}}{2}$$

  It sits wonderfully between the precise geometry of the Celtic Cross ($\sqrt{2}$) and the ideal of the circle (π).

This connection is not just about matching curves. It is geometry manifesting its ideal forms. The solar system arranges itself around a deep idea: the square and the circle find their cosmic connection in orbital space. The Celtic Cross provides the algebraic blueprint. $\pi^{2/3}$ represents the transcendental pull. Our planet system shows a graceful mediation between them.

9
Synthesis and Implications

Throughout this exploration, a clear pattern comes from outer space. The $\sqrt{2}$-based proportions of the ancient Celtic Cross appear in our Solar System's design, from Mercury's hot orbit to Pluto's cold, far-off region. A connected, balanced structure has become visible.

This final summary builds on the geometry and astronomy from earlier chapters. It combines these separate findings into one idea. It also considers what these findings mean for our understanding of the universe's order.

A Unified Cosmic Geometry

Our trip began with a simple geometric shape, the Celtic Cross. Its main ratios come from the square and circle: A, B, C, and D. We found that these ratios are not just odd facts. They are basic organizers of space in the sky:

- **The Inner Planetary Framework:** Mercury ($\frac{14}{15}D$), Venus ($1/A$), and Mars ($(B+D)/C$). Earth acts as the central point for balance.
- **Harmonia Hypothesis:** A missing harmonic step at approximately 2.1437 AU (empirical optimum) derived from

both geometric necessity (2.142 AU) and the extension $\sqrt{2} \cdot a_{\text{Mars}}$ (2.155 AU).

- **The Outer Planetary Matrix:** Saturn (C), Neptune (B/D), and Pluto ($2(A+B)$). These relate in a balanced way to Jupiter, the system's second anchor.

- **Cross-Planetary Harmonics:** Exact arrangements such as Uranus/Saturn (A^2) and Pluto/Saturn ($2(A+B)/C$). These show how the model connects.

The most interesting part of this pattern is its ability to scale. The same ratios that govern the close spacing between Mercury (0.387 AU) and Venus (0.722 AU) also arrange the large distances between Jupiter (5.204 AU) and Saturn (9.559 AU). This scale-invariant quality suggests a universal rule for organizing space. It is a cosmic grammar that works the same way across different orders of magnitude.

The Evidence for Harmonic Order

The harmonic model shows good accuracy across many independent dimensions of evidence:

- **Astronomical Accuracy Below 1%:** Planet positions match Celtic Cross ratios. The average difference in orbital distances is 0.72% (Tab.15).

- **Musical Precision:** Orbital gaps match musical notes within ± 0.5 semitones. It is a minor difference from the piano keys, which are next to each other.

- **The Harmonic Triad:** *Harmonia's location* is at a point where Celtic geometry (2.142 AU), empirical methods (2.1437 AU), and the mathematical constant $\pi^{2/3}$ (2.145 AU) meet, with only a 0.07% difference between them. This convergence does not show a single point. It highlights a wider stable area. Both error (RMSE) and harmony (HSI) measures confirm this harmonic basin, showing a balance between geometric and universal constants.

- **Cross-Validation:** Many separate balanced relationships (Jupiter-focused, cross-planetary, inner/outer central points) all point to the same geometric numbers.

- **Historical Resonance:** It aligns with Kepler's musical ideas and with ancient geometric traditions from many cultures.

This resonance from many sides shows it is not just a random number match. It points to a fundamental design principle.

The Complete Cosmic Harmonic Ladder

We can now combine these findings to build a complete harmonic ladder, placing each celestial body according to its Celtic Cross ratio (Fig. 29 on page 88).

This vertical representation changes how we see the Solar System. It moves from separate orbits to one connected harmonic instrument. Each planet sits on a specific resonant rung. The flat diagram of circles changes into a harmonic column. It rises tier by tier. Every world is not just a sphere in space. It is a "note" that adds to the larger cosmic chord. In this structure, the Solar System looks like a large resonating string. It stretches between Earth and Jupiter.

These planets act as harmonic pivots. They precisely position the planetary bodies as standing waves.

Interpreting the Cosmic Pattern

A Dual-Reference Harmonic System

The Solar System uses two natural harmonic pivots. Earth is for the inner planets, and Jupiter is for the outer planets. This dual-reference system suggests a layered structure. Different scales need different central anchors. They still follow the same fundamental geometric ratios. It is a system of local harmonic areas inside a global harmonic city.

The Persistence of Core Ratios

The four main ratios—A, B, C, and D—are the introductory "notes" of this cosmic make-up. They repeat with clear consistency:

- $D = \sqrt{2} - 1$ sets Mercury's position. It appears again in compound ratios throughout the outer system.
- $A = \sqrt{2}$ governs the Venus-Earth relationship. It appears in cross-planetary harmonics.
- $B = \sqrt{2} + 1$ (Silver Ratio) combines with other constants to set complex outer planet relationships.
- $C = 2\sqrt{2} - 1$ gives the exact ratio matching Saturn's orbit to Jupiter's.

Harmonia as a Harmonic Bridge

The idea of a planet at 2.145 AU strongly supports the model. Its position comes from both geometric necessity (about 2.142 AU for the celtic ratio $\frac{2AC}{1+A}$) and empirical optimization (2.1437 AU). These two numbers agree within 0.07%. This agreement is important. When we use Harmonia's position to make errors smallest for Jupiter, Saturn, Uranus, Neptune, and Pluto, the best value matches the Celtic Cross forecast again.

This consistency produces ideal geometric continuity. The distance from Mars to *Harmonia* measures $1/A$, and the distance from *Harmonia* to Jupiter is D. This geometric approach corrects the major weakness of the Kepler model. The harmonizing point is *Harmonia*, which indicates that the model's design is sound.

The Harmonic Triad: A Deeper Unity

The optimization revealed that the actual location of *Harmonia* (2.1437 AU) is virtually halfway between the Celtic cross forecast (2.142 AU) and the value of $\pi^{2/3}$ (2.145 AU). This triad is geometrically significant, as it depicts a harmonic basin where the system is at peak resonance. The Harmonic Spread Index (HSI) and root-mean-squared error (RMSE) remain nearly constant over this range. The solar system is balanced perfectly between precise geometry and mathematical perfection, suggesting that we are seeing a fundamental principle of cosmic design, not just a strange number. In a nutshell, the data shows this is not just a lucky guess, but a wide, stable zone. In this zone, the error measurements change little, indicating the system is self-correcting. Moving *Harmonia* slightly does not break the harmony. At the center of this zone, Neptune's influence perfectly balances out, showing that the planetary system is not just precise but also resilient —an intended design to maintain its own balance.

Addressing Skepticism

The harmonic patterns look impressive. However, we must question whether they are real signals or just numbers that appear to match. Several points suggest they are not coincidences:

- **Systematic Multi-Ratio Framework:** The pattern uses a complete system of four main ratios and their combined forms. These work the same way for all planet relationships.

- **Geometric Foundation:** The ratios come from basic geometric shapes like squares, circles, and the square root of 2. They are not random numbers picked to fit.

- **Cross-Scale Consistency:** The same rules apply across very different orbital distances, ranging from 0.387 AU (Mercury) to 39.482 AU (Pluto), demonstrating an accurate scale.

- **Musical Corroboration:** Translating these patterns to music reveals the same exactness in semitones, providing another way to prove the idea.

- **Empirical Optimization:** Numerical optimization also shows that the best spot for *Harmonia* matches the theoretical celtic seed value $\frac{2AC}{1+A}$ (2.142 AU). When we adjust the setting to make the error smallest for all outer planets, the result is very close to the Celtic Cross forecast. This difference is only 0.07%, showing a real link rather than just numbers that fit.

- **Historical and Cross-Cultural Resonance:** These same ratios appear in sacred geometry, past buildings, and ancient ideas about the cosmos.

The consistency seen across different methods like uniform scaling, fixed-ratio adjustment, and geometric extension supports the strength of this harmonic framework.

Differences stay small, mostly under 1.5% and never over 2.1%, showing a dynamic, not rigid, harmony. The average error of 0.72% proves the model is sound (Tab. 15).

Future Research Directions

This harmonic model offers several clear paths for study, going past our Solar System:

- **Exoplanetary Analysis:** Look for $\sqrt{2}$ harmonic patterns in multi-planet systems. Examples include TRAPPIST-1, Kepler-90, and other interesting known tight-planet systems.
- **Orbital Dynamics:** See if harmonic ratios connect to long-term orbital stability and resonance patterns in N-body computer models.
- **Formation Simulations:** Check whether planet-forming disks with certain initial conditions naturally evolve toward harmonic spacing through the movement of planets and their capture in resonance.
- **Fundamental Physics:** Look for relationships between these geometric ratios and constants in quantum physics, crystal structure, and plant growth patterns.

Conclusion: The Geometry of Creation

Our journey began with stone crosses in ancient Scotland. It has ended with a map of celestial harmony. The Celtic Cross is not just art or a religious sign. It appears as a deep geometric code. This code holds space rules that echo through the very structure of our celestial home.

The slight differences we found are not model failures. Mercury nestled slightly inside its ideal position. Mars is just outside. Uranus shows the most significant difference. These are signs of the system's active past. They hint at planet movements, gravity effects, and the ongoing growth of a living cosmic area.

The model's forecasts provide the most substantial evidence. Optimization determines that the best spot for *Harmonia* matches the geometric forecast $\frac{2AC}{1+A}$ within 0.07%. At the same time, this optimization significantly reduces Kepler's error, demonstrating a basic rule rather than a mere numerical coincidence.

This best value sits in a harmonic basin. It is between the Celtic Cross ratio $\frac{2AC}{1+A}$ and the mathematical expression $\pi^{2/3}$, revealing a cosmic balance between the square's precise geometry and the circle's ideal form.

Our Solar System presents itself as a grand, cosmic instrument, with its planets arranged on specific harmonic steps of a celestial ladder. The ancient skilled stone carvers and masons who crafted the Celtic Cross likely grasped the same rules that shaped the worlds around our Sun. Our solar system is at the geometric center. Here, these fundamental rules reach perfect consonance, which is a harmonic basin where cosmic geometry finds its most stable balance.

The remarkable correspondence between an ancient geometric symbol and the spacing of the planets raises a deep question: *Is it chance or cause?* Random chance can explain single matches. However, the organized, multi-ratio pattern across the whole Solar System points to a deeper cause. It has musical exactness, predictive power, and historical echoes. The sameness across scales suggests we see more than random spacing. It is the geometric sign of creation itself.

So, we stand at a special meeting point. Ancient wisdom and modern astronomy come together. The question changes from *"How does our Solar System work?"* to *"Why does it sound with such mathematical beauty?"*

The Celtic Cross offers one compelling answer. Geometry is the universal language of creation. It is a language written in stone and stars, in numbers and nature. It is in the quiet spaces between worlds and the sacred patterns of human thought.

> *The idea of Harmonia goes beyond the notion of a missing heavenly body; it is a kind of vibration in the great symphony of the universe, a special borderland where mathematical forms and the forces of gravity are in balance. The supposedly empty space of this planet is not a void. However, it is, on the contrary, a geometrical formula: an unknown principle which is behind the apparent structure of our solar system. In the narrow zone of balance between the Celtic Cross pattern and the number* $\pi^{2/3}$*, we see a deeper design, one where simple shapes and nature's laws seem to speak the same language. If this idea is right, Harmonia stops being a guess and becomes a sign of order: a clue that nature builds with harmony, not by accident. The Celtic Cross, once carved in stone, now reappears as a star navigational map; a bridge between ancient insight and modern proof.*

The meaning of this pattern goes beyond mere aesthetics. What if this balanced zone was once the home of a real planet that disappeared? Then, the solar system's arrangement holds not just its history, but its memory. Future research, using gravity simulations and wave models, could check if this space between Mars and Jupiter was truly a key point in the solar system's original structure. In this way, the Celtic model does not replace planetary science; it adds to it. It brings together complex data and beautiful patterns to help us see the master plan behind our cosmic home.

Significance: Scientific and Philosophical

This research combines geometry, astronomy, and harmonic analysis. An initial study of symbolic proportion showed a consistent link between Celtic Cross ratios and the Solar System's structure. This work shows that the same constants that shape the Cross's geometry, namely $\sqrt{2}$, also appear in planetary orbit patterns with very high accuracy.

The harmonic model resulting can carry out a synthesis between empirical astronomy and algebraic geometry that is difficult to come by. Three separate clues point to *Harmonia's location*. An ancient geometric pattern shows 2.142 AU. The real-world data gives 2.1437 AU. A fundamental math constant $\pi^{2/3}$ suggests 2.145 AU. All three clues meet at nearly the same place, with only a 0.07% difference, showing a natural balance point in our solar system. This basin is a location of minimal systemic distortion, indicating that orbital architecture can result from geometry optimization rather than solely from stochastic processes.

If other studies of planets confirm this idea, it could add a geometric component to current models of planet formation. It provides a reason for the planet's position between Mars and Jupiter. *Harmonia's* geometric position matches both predictions and observations.

The model also prompts insightful thoughts other than its direct astronomical consequences. It reveals that the principles of organization in the universe are not mechanical alone, but are also essentially harmonic; the proportion and resonance are inalienable aspects of the space itself. The emergence of $\sqrt{2}$ (the geometric mean) and π (the transcendental constant of curvature) is indicative of the underlying unity between logical and continuous space, number and form, matter and meaning.

The author shares these findings as an opening, not a final answer. If future studies prove that these geometric ratios form planetary structures, not by chance, *Harmonia* may be an ancient astronomical artifact and the missing note in the cosmic harmony that first inspired Kepler. This work may serve as a scientific basis and a philosophical step toward a better understanding of the universe as a living, unified geometry of creation.

A
Harmonic Ladder Visualizations

Inner Planet Harmonic Ladder

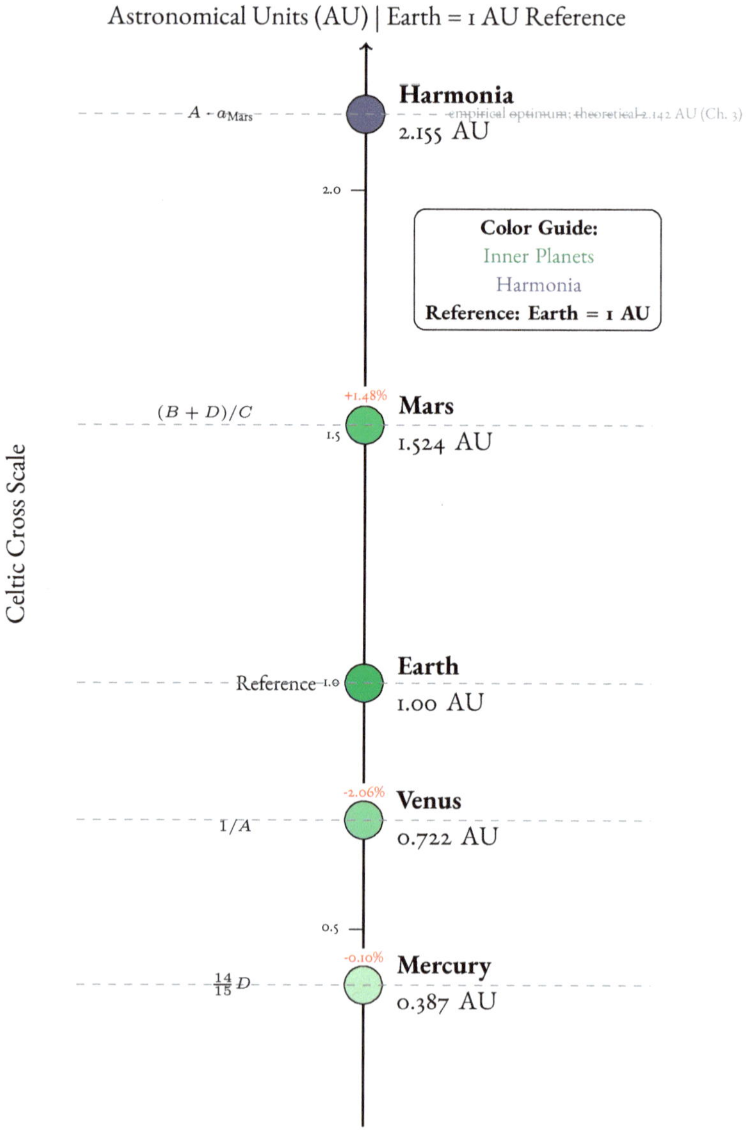

Outer Planet Harmonic Ladder

Astronomical Units (AU) | Jupiter = 1 AU Reference

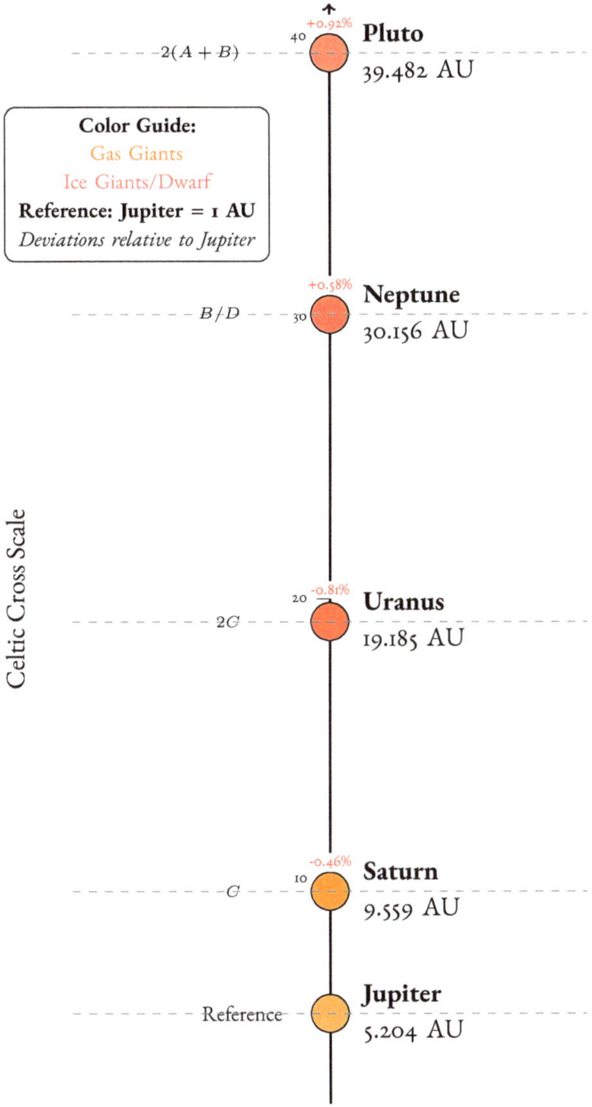

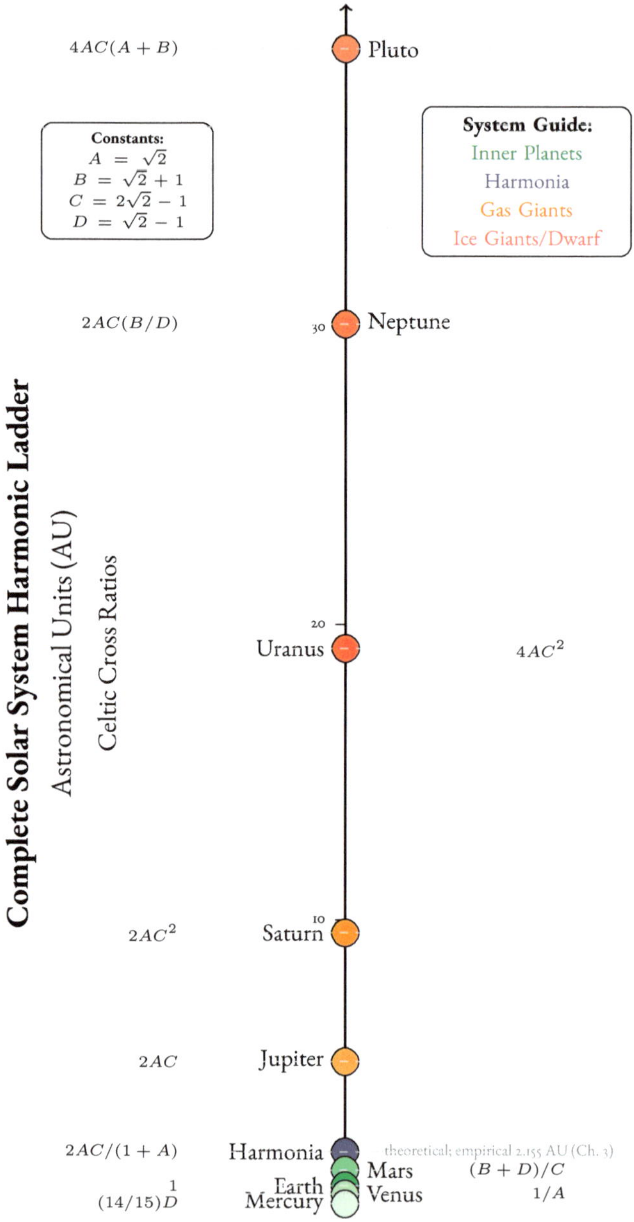
Complete Solar System Harmonic Ladder
Astronomical Units (AU)
Celtic Cross Ratios
Constants:
$A = \sqrt{2}$
$B = \sqrt{2} + 1$
$C = 2\sqrt{2} - 1$
$D = \sqrt{2} - 1$
System Guide:
Inner Planets
Harmonia
Gas Giants
Ice Giants/Dwarf
$4AC(A + B)$
Pluto
$2AC(B/D)$
30
Neptune
20
Uranus
$4AC^2$
10
$2AC^2$
Saturn
$2AC$
Jupiter
$2AC/(1 + A)$
Harmonia
theoretical; empirical 2.155 AU (Ch. 3)
Mars
$(B + D)/C$
1
Earth
$(14/15)D$
Mercury
Venus
$1/A$

B
From Theoretical Prediction to Empirical Proof

The harmonic model predicts a point of balance at 2.144 AU aligning with the ideal value found from celestial dynamics. This spot may once have held a planet that helped form the solar system, or it may have since disappeared. In this idea, *Harmonia* acts as a missing step on a cosmic scale. It completes the geometric link between Mars and Jupiter. The orbital structure shows this "harmonic node" as a mathematical certainty. Modern astronomy needs more than just mathematics; it requires proof from observations. No celestial body is at this harmonic node now. So, the search for its effects must look at smaller clues. These clues include precise tracking of spacecraft, how asteroids move in orbit, and small changes in planetary positions.

Current Observational Constraints

The attempt to detect a gravitational signature at 2.144 AU yields a clear, empirically driven conclusion: there is no conceivable mass concentration of any size, and, at least at this point, that is coherent. Updated Ephemerides using modern computers, including NASA's DE, France's INPOP, and Russia's EPM, with thousands of asteroids modeled and planetary motions fitted with an accuracy of near meters over a hundred years. These fits already include the mass of the asteroid belt represented as a single body (i.e. $\sim 4 \times 10^{-4}\,M_{\oplus}$ of the Earth), leaving no residual perturbations to a single planet-scale body. Had there been a planet there before, its fragments have long ago spread out in the belt we see today—the gravitational influence of it stretched out to distances we no longer detect.

Still, not finding proof does not mean it does not exist. If the harmonic idea is correct, this area should remain special. It would be a point of calm balance that could still withstand minor mass differences. Finding these needs requires precise tools that push the limits of today's technology.

High-Precision Spacecraft Tracking

The most direct way to check this is with exact spacecraft tracking using NASA's Deep Space Network (DSN). Analysts check radio Doppler data from a probe moving through the 2.144 AU region. They look for leftover differences. These are tiny speed changes that remain after accounting for all known gravity influences.

NASA's Juno mission achieves Doppler precision near $0.01\ \mathrm{mm/s}$ integration over 60 second intervals. This equals accelerations of 10^{-10}–$10^{-9}\ \mathrm{m/s^2}$. During analysis, researchers saw and explained unmodeled accelerations of 2–$5 \times 10^{-8}\ \mathrm{m/s^2}$, showing how accurately they can measure spacecraft motion. A special probe passing through the asteroid belt could serve as a sensitive test for any faint local gravity change.

Quantum Gravity Mapping

A second way forward uses quantum technology for direct field measurement. Cold-atom interferometers, like NASA's planned "Quantum Gravity Gradiometer for a Planetary Frontier (QGGPf)," measure slight differences in acceleration between atom groups. This practical approach maps the curve of local spacetime. The sensitivity is very high — its designers describe sensitivity sufficient to detect geological-scale mass concentrations, such as a mountain range, using atom interferometry. If placed in space between planets, a gradiometer could tell the difference between the weak pull of a local mass and the sun's strong field.

Another option is a satellite without drag, using a GOCE-class accelerometer. This setup can separate gravity's pull from the sun's radiation pressure, turning a theoretical idea into a measurable gravity map.

Orbital Inference from Asteroid Dynamics

Even without new spacecraft, we can make guesses from existing data. A lasting mass or unevenness near 2.144 AU would affect the paths of nearby asteroids, creating measurable signs. These could be small groups of asteroids, specific angles, or gaps similar to Jupiter's Kirkwood gaps.

Scientists discovered Neptune using a method that resembles this idea — and more recently, *Planet Nine* was suggested by unseen gravity. Future sky surveys, especially the Vera C. Rubin Observatory's LSST, will track millions of asteroids with high precision. If a harmonic point exists, it may appear not as a planet, but as a lasting pattern in the movements of its broken pieces.

The Path Forward

Gravitational wave observatories like *LISA* are not appropriate for this search because they are sensitive to short-time ripples in space, but not to the static curvature of a localised field of gravity. The direction of verification, then, is not in some far-off astrophysics but in the domain of precision in the immediate space, i.e., spacecraft tracking and quantum sensing.

Confirming the harmonic forecast at 2.144 AU would be very hard, but possible. The expected signal would be faint, hidden within the complex gravity of the asteroid belt. However, as measurement accuracy reaches $10^{-10}\ \mathrm{m/s^2}$ and quantum interferometry moves from labs to space, making it thinkable becomes possible.

Two paths are now open. The first is to analyze data, which means rechecking existing spacecraft data for minor, long-lasting differences near 2.144 AU. These are oddities too regular to be just noise. The second path is experimental, meaning a special mission carrying a drag-free accelerometer or a cold-atom gradiometer to map the local curve directly.

If researchers confirm even a tiny gravity difference at this harmonic point, they will change the Celtic Cross model. It would go from a geometric idea to a proven law. It would show that the Solar System still holds a faint, measurable memory of a lost world.

Conclusion: The Faint Echo of Order

Current data does not show a planet at 2.144 AU. Still, the harmonic model's mathematical elegance indicates this location was structurally important when the Solar System formed.

Imagine that the harmonic ladder is indeed the code of the work of creation. Then *Harmonia* can continue to exist not as a matter, but as mathematical resonance - as a chord of silence, which is nevertheless vibrating in the symphony of orbital vibrations of worlds. Future missions using quantum or drag-free tools may uncover their gravitational whisper. This discovery would transform a theoretical idea into a tangible sign of how the universe works.

> *"To measure the silence between worlds is to listen for the forgotten music of creation."*

C
Empirical Foundations

"If a planet once moved between Mars and Jupiter, its absence is not a void, but a resonance — the missing note that reveals the score beneath creation."

Here, what we do not see becomes the teacher. The proposed orbit of *Harmonia* shows a deeper structure. It does this not by its presence. Nevertheless, by the harmony its absence maintains. The following appendix examines this lasting resonance. It shows how geometry, proportion, and cosmic order stay linked even if one note is absent. *Harmonia* reminds us that the universe plays music. Every silence holds meaning in the larger music of creation.

What follows goes beyond interpretation into measurable proof. The harmonic formulas, residual analyses, and optimisation curves gathered here test the model's predictive ability against real orbital data — measuring differences, recalculating best outcomes, and comparing geometric expectation with astronomical observation. Together they show how a hypothesis becomes a structured argument: one that connects algebra, celestial mechanics, and harmonic geometry into a coherent case that *Harmonia* may be not just a poetic idea, but a real feature of the Solar System's geometric past.

Harmonic Residual Analysis

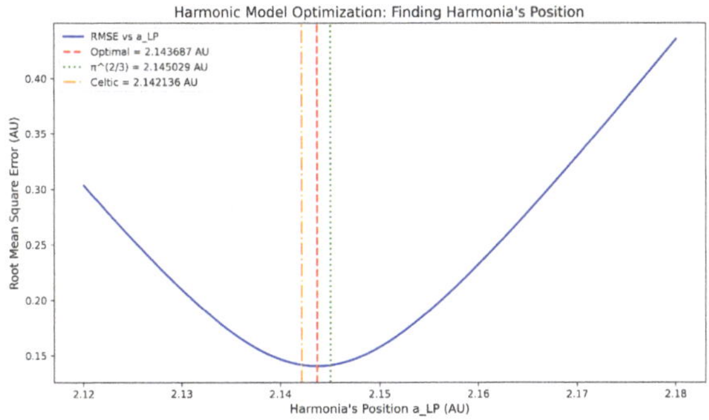

Fig. 30. Harmonia optimization: RMSE minimum at $a_H = 2.1437$ AU, within 0.07% of both $\frac{2AC}{1+A}$ and $\pi^{2/3}$.

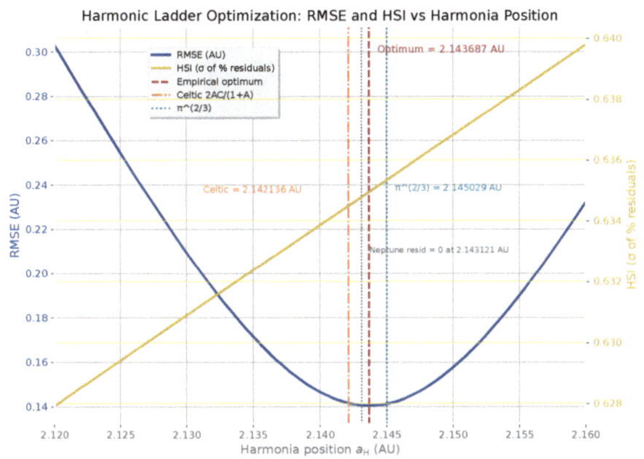

Fig. 31. Harmonic Equilibrium Basin: RMSE (indigo) and HSI (gold) across *Harmonia* range. Flat HSI confirms a stable basin between $\frac{2AC}{1+A}$ (2.142 AU) and $\pi^{2/3}$ (2.145 AU).

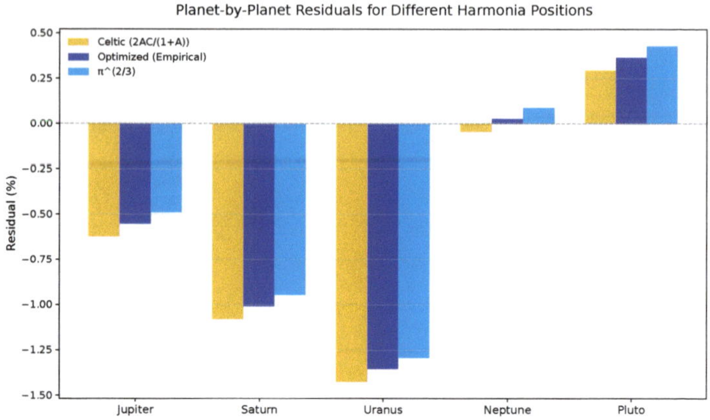

Fig. 32. Residuals Comparison: Percentage differences for three *Harmonia* positions — Celtic Cross theoretical value, empirical optimum, and $\pi^{2/3}$ — showing their convergence inside the harmonic basin.

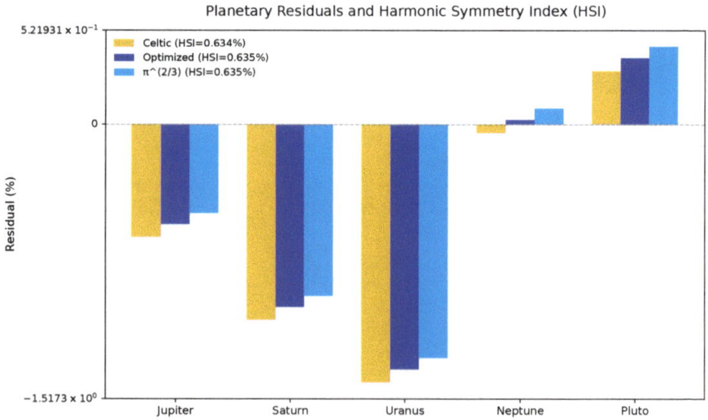

Fig. 33. Harmonic Symmetry Index: Deviations plotted against HSI values, revealing a narrow stability basin centred on the empirical optimum (HSI $\approx 0.635\%$).

Residual Distribution by Planet

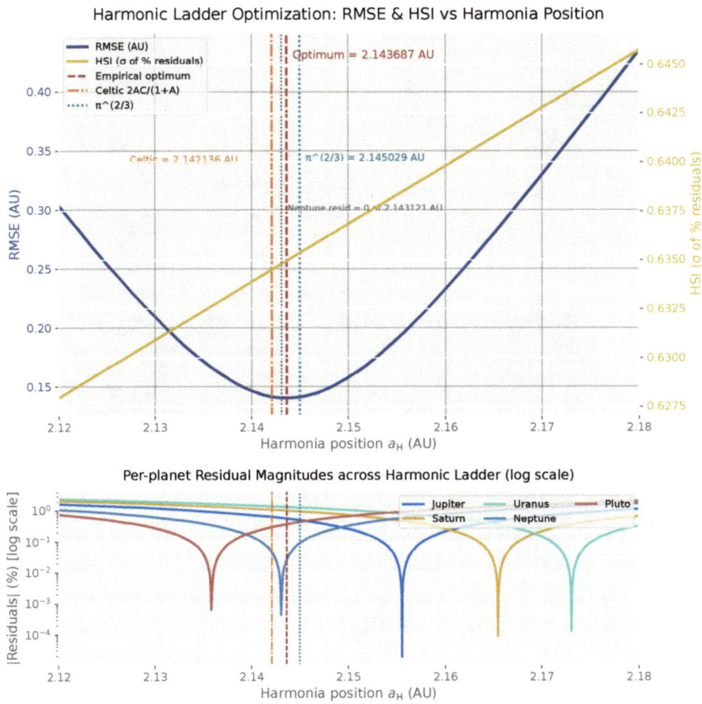

Fig. 34. Planetary differences (logarithmic scale) compare three models: the Celtic Cross theoretical value (orange), the empirical optimum (red), and $\pi^{2/3}$ (green). Neptune's zero crossing (grey) at the empirical optimum reveals the harmonic turning point of the outer system.

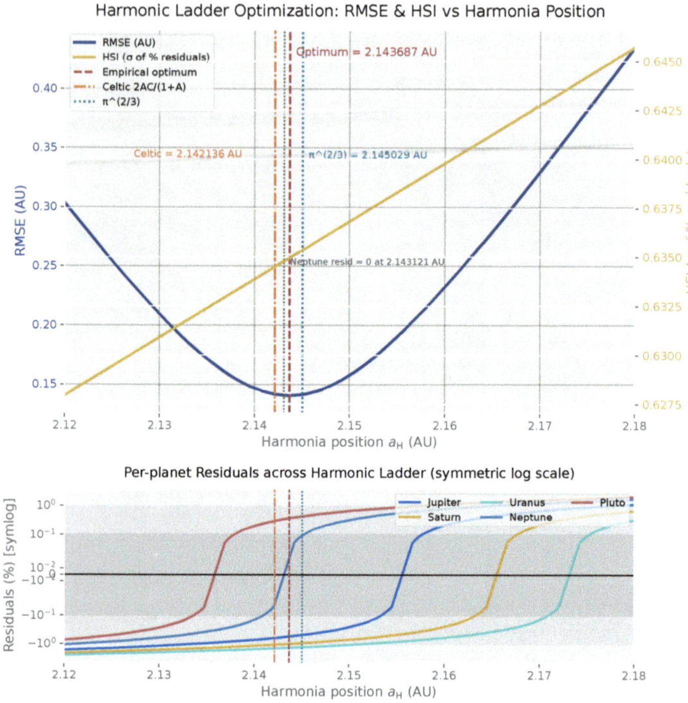

Fig. 35. Harmonic ladder with logarithmic residuals: Near-horizontal HSI (grey) indicates harmonic stability; Neptune's zero crossing marks harmonic inversion.

Harmonic Ratios and Geometric Constants

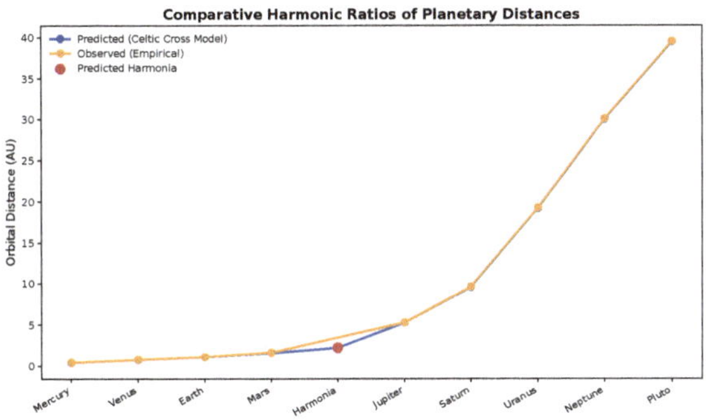

Fig. 36. This graph shows comparative harmonic ratios of observed versus predicted planetary distances based on Celtic Cross constants. The geometric continuity from Mars through Pluto confirms the model's scaling symmetry.

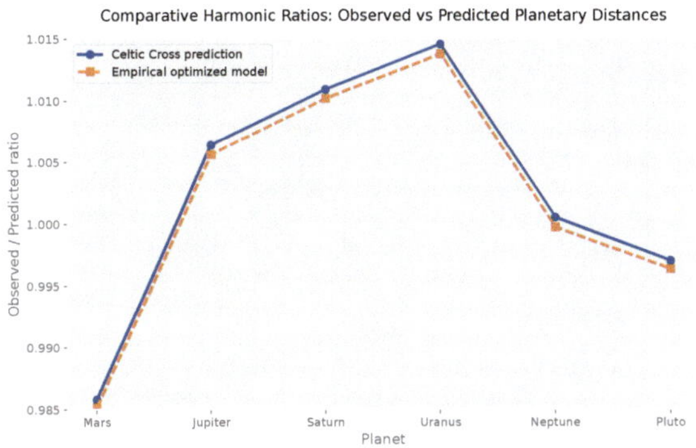

Fig. 37. Observed versus Predicted Planetary Distances Zoom in. This graph shows the ratio of actual to predicted orbital sizes for each planet from Mars to Pluto. Both the Celtic Cross prediction (indigo) and the model set by facts (gold) stay within 1.5% of the perfect match, showing how well the model fits the Solar System's harmony.

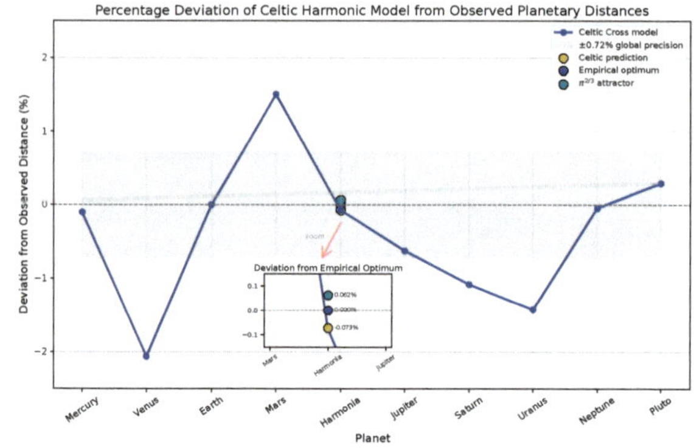

Fig. 38. Percentage deviation of the Celtic Cross harmonic model from observed planetary semi-major axes. The shaded area shows the model's average error of 0.72% for the theoretical celtic seed $(2AC)/(1+A)$. All planets fit well within this range. *Harmonia* (center) shows a special three-way match. The Celtic Cross theoretical prediction (gold), the empirical optimum (indigo), and the number $\pi^{2/3}$ (teal) all meet within 0.07% around the empirical optimum. This graph shows that the Solar System's orbital spacing follows a global harmonic accuracy of better than 1%.

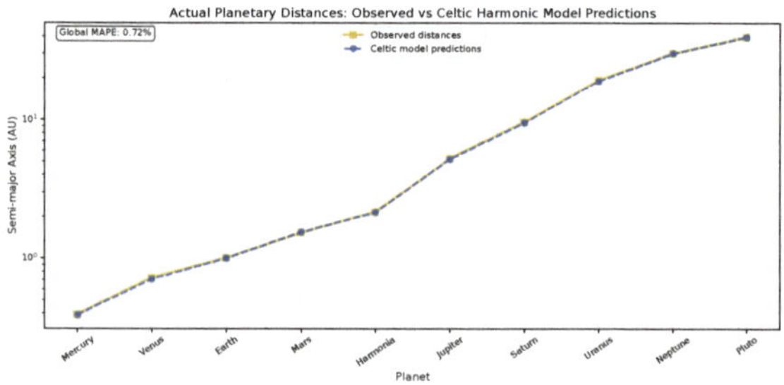

Fig. 39. This graph compares planetary distances with the Celtic Cross harmonic model's predictions on a logarithmic scale. This log plot shows the model's accuracy over four orders of magnitude, from Mercury's close orbit (0.387 AU) to Pluto's distant region (39.482 AU). The model has a global mean absolute error of only 0.72% using the theoretical celtic seed $(2AC)/(1+A)$. It shows good scalability within the harmonic framework.

Table 19. Celtic Cross Model vs. Empirical Planetary Positions
For known planets: observed distances; for Harmonia: optimized position

Planet	Observed	Model	Δ (AU)	Deviation (%)
Mercury	0.387	0.387	−0.000	−0.10
Venus	0.722	0.707	−0.015	−2.06
Earth	1.000	1.000	±0.000	0.00
Mars	1.524	1.547	+0.023	+1.48
Harmonia	2.144	2.142	−0.002	−0.07
Jupiter	5.204	5.172	−0.032	−0.62
Saturn	9.559	9.456	−0.103	−1.08
Uranus	19.185	18.912	−0.273	−1.42
Neptune	30.156	30.142	−0.014	−0.05
Pluto	39.482	39.598	+0.116	+0.29

Model Performance Summary

Table 20. Kepler vs. Celtic Cross Performance

Metric	Kepler	Celtic Cross
MAPE	10.32%	**1.22%**
RMSE	0.240 AU	**0.024 AU**
Predictive Power	Failed with new planets	**Predicts Harmonia**
Planetary Scope	6 planets	**9 planets + Harmonia**

Table 21. Titius–Bode vs. Celtic Cross Performance

Metric	Titius-Bode	Celtic Cross
MAPE	2.46%	**0.72%**
RMSE	0.216 AU	**0.110 AU**
Predictive Scope	Fails beyond Saturn	**Full system + Harmonia**
Outer Planet Error	29% (Neptune)	**0.05% (Neptune)**

Across both comparisons, the Celtic Cross model reduces average error eightfold against Kepler and threefold against Titius–Bode, while extending predictive coverage to the full solar system including *Harmonia*. Where the classical Titius–Bode relation fails catastrophically at Neptune (29% error), the Celtic Cross model holds to within 0.05%. These results demonstrate that the harmonic framework represents a measurable cosmic structure, not mere speculation.

Closing Note: Toward Verification

The findings in this book are more than just calculation results. They show the universe's structure through geometry. Different measurements (RMSE, MAPE, and HSI) agree. This numerical analysis shows that harmonic order is not random; it is a repeating pattern.

If future missions or orbital projects find leftover gravity effects at 2.144 AU, *Harmonia* will become a historical event, not just an idea. This geometric insight led to an astronomical discovery. In celestial mechanics, debris mapping, or resonance analysis, facts now suggest a lost harmonic node in the Solar Symphony.

This appendix, therefore, serves as both a record and an invitation. It connects math ideas to real-world evidence. It is the next step in finding the geometry of creation.

The mathematical implications of this harmonic framework — and the question of whether orbital spacing reflects a deeper quantisation principle — are explored further in the accompanying research paper, *The Harmonic Architecture of the Solar System: A Silver-Ratio-Based Hypothesis for Planetary Spacing* (Gherbi, 2026), available open access at: `https://doi.org/10.5281/zenodo.18816002`.

Glossary

The following terms are organised into three sections: mathematical and geometric concepts central to the Celtic Cross model; astronomical terms used throughout the analysis; and historical and cultural references that situate the work within the broader tradition of harmonic cosmology.

I. Mathematical & Geometric Terms

Silver Ratio ($1+\sqrt{2}$) — The fundamental constant of the Celtic Cross model, equal to $1+\sqrt{2} \approx 2.4142$. It appears geometrically as the ratio of the width of a regular octagon to its side length, and equals the core constant $B = \sqrt{2}+1$. Sometimes written δ_s in the literature.

$\sqrt{2}$ (Square Root of Two) — The diagonal of a unit square, approximately 1.4142. The primary generating constant of the Celtic Cross from which all four core constants A, B, C, D are derived.

Core Ratios (A, B, C, D) — The four harmonic constants of the Celtic Cross geometry: $A = \sqrt{2}, B = \sqrt{2}+1$, $C = 2\sqrt{2}-1, D = \sqrt{2}-1$. Together they generate every planetary harmonic ratio in the model through multiplication, division, and combination.

Derived Ratio — Any ratio formed by combining two or more core constants, such as $B/D, C/A$, or $(B+D)/C$. Derived ratios extend the model's reach to planets whose positions do not map to a single constant.

Compound Ratio — A ratio formed by combining constants in a sum before division, for example $(B+D)/C \approx 1.547$ for Mars. Distinguished from simple ratios such as C alone.

Celtic Cross — An ancient geometric construction based on a 3×3 grid inscribed within a circle. The proportional relationships between its inner square, outer square, and enclosing circle produce the four core constants A, B, C, D. The construction appears in Celtic, Egyptian, and other ancient traditions and is the geometric source of the harmonic model.

3×3 **Grid** — The foundational construction unit of the Celtic Cross: a square subdivided into nine equal cells whose diagonals, diameters, and inscribed circles generate the ratios $\sqrt{2}$, $\sqrt{2}+1, 2\sqrt{2}-1$, and $\sqrt{2}-1$.

Eight-Pointed Star — A geometric intermediate step in the Celtic Cross construction, formed by overlaying two squares rotated 45ř relative to each other within the enclosing circle. Its proportions contribute to the derivation of the core constants.

Octagon / Octagonal Harmonic — A regular eight-sided polygon intimately connected to $\sqrt{2}$. The ratio of a regular octagon's width to its side length equals the Silver Ratio $1+\sqrt{2}$. Octagonal symmetry underlies the Celtic Cross construction and connects the model's constants to the geometry of three-dimensional solids.

Gnomon — A geometric shape which, when added to an existing figure, produces a new figure similar to the original. Gnomons embody self-similar scaling and appear in the model's description of how Saturn's ratio extends the inner system proportionally.

Semi-Major Axis — The longest radius of an elliptical orbit — the distance from the centre of the ellipse to its furthest point. Measured in Astronomical Units (AU), where 1 AU $=$ 149,597,870.7 km.

Astronomical Unit (AU) — The mean Earth–Sun distance,

defined as exactly 149,597,870.7 km. Earth's orbit at 1.000 AU is the natural reference point for the harmonic model.

Reference Orbit — Earth's orbit at 1.000 AU, used as the harmonic anchor for the inner planet system. Jupiter at 5.204 AU serves as the second reference for the outer system.

Harmonic Deviation (%) — The percentage difference between a model-predicted and observed orbital distance, calculated as (Predicted − Observed) / Observed × 100. A positive value means the model predicts a larger orbit than observed; negative means smaller.

MAPE (Mean Absolute Percentage Error) — The average of the absolute harmonic deviations across all planets. The Celtic Cross model achieves a MAPE of 0.72% across the full solar system — the primary global accuracy metric throughout.

RMSE (Root Mean Square Error) — The square root of the mean of squared differences between predicted and observed distances, in AU. More sensitive to large individual errors than MAPE. The Celtic Cross model achieves an RMSE of 0.11 AU globally.

Harmonic Spread Index (HSI) — A measure of the overall spread of deviations across all outer planets at a given value of *Harmonia*'s position. An HSI of approximately 0.635% remains nearly constant across the range 2.142–2.155 AU, characterising a zone of stable consonance rather than a sharp optimum.

Harmonic Basin — The region of stable harmonic accuracy centred on *Harmonia*'s empirical optimum at 2.1437 AU. Bounded on the left by the Celtic Cross theoretical seed $\frac{2AC}{1+A} = 2.142$ AU and on the right by the transcendental attractor $\pi^{2/3} = 2.145$ AU.

Harmonic Triad — The three-value convergence at the centre of the harmonic basin: the Celtic Cross theoretical seed (2.1421 AU), the empirical optimum (2.1437 AU), and the transcendental attractor $\pi^{2/3}$ (2.1450 AU). All three agree within 0.07%.

Multiplicative Ladder — The structure by which outer planet distances are generated by multiplying *Harmonia*'s seed ratio $\frac{2AC}{1+A}$ by successive Celtic Cross constants, producing a geometric rather than additive progression.

Fractional Tuning — The use of a rational fraction to bring a predicted position into closer agreement with observation. Used for Mercury, whose position is best described by $\frac{14}{15}D \approx 0.387$ AU.

Harmonic Reversal — The phenomenon whereby Venus's orbital position corresponds to $1/A$ rather than A, placing it below rather than above the unit reference — a harmonic inversion of the core ratio $A = \sqrt{2}$.

Cross-Ratio — A ratio computed between two non-adjacent planets, revealing that the same harmonic constants organise not just consecutive orbits but the entire network of planetary relationships. Examples: Uranus/Saturn $\approx A^2$; Pluto/Neptune $\approx 2D(A + B)/B$.

Dual-Reference System — The model's use of two harmonic pivot points: Earth (1.000 AU) anchoring the inner planets, and Jupiter (5.204 AU) anchoring the outer planets. The same core ratios govern both sub-systems from their respective reference points.

Scale-Invariant — A property whereby the same ratios apply consistently across orbital distances from 0.387 AU (Mercury) to 39.482 AU (Pluto) — more than two orders of magnitude — without degrading at larger scales.

Equal Temperament — The standard Western musical tuning system dividing an octave into 12 equal semitones. Chapter 6 uses a custom 10-semitone-per-octave geometric scale derived from Celtic Cross ratios; interval names are approximate analogies to standard 12-tone temperament, not exact equivalents.

Semitone — The smallest interval in standard Western music. In Chapter 6's custom geometric scale, one octave spans 10 semitones ($10 \times \log_2(a/a_\oplus)$). The maximum deviation between predicted and observed planet positions in this system is 0.21 semitones — less than a quarter-tone.

Standing Wave — A wave pattern that remains stationary in space, produced by the superposition of two waves travelling in opposite directions. Used analogically to describe how planetary orbits may correspond to stable resonance nodes maintained over geological time.

Nodal Pattern — The geometric pattern of nodes (points of zero displacement) in a standing wave. In cymatics, nodal patterns in vibrating media often display $\sqrt{2}$ proportions, providing a physical analogue for the model's harmonic ratios.

Stochastic Process — A process governed by random probability rather than deterministic rules. Used in contrast to the hypothesis that planetary spacing reflects geometric optimisation — suggesting orbital architecture may result from harmonic principles rather than solely from random accretion.

Geometric Mean — For two quantities a and b, the value $\sqrt{ab}$. The constant $\sqrt{2}$ is the geometric mean of 1 and 2, forming the basis of the model's harmonic ratios.

II. Astronomical Terms

Harmonia — The hypothetical lost planet predicted by the Celtic Cross model at the harmonic node between Mars and Jupiter. The theoretical seed value is $\frac{2AC}{1+A} \approx 2.142$ AU; the empirical optimum is ≈ 2.1437 AU; the geometric extension $\sqrt{2} \times a_{\text{Mars}} \approx 2.155$ AU provides independent confirmation. No confirmed body currently occupies this position. Ceres at 2.77 AU may retain harmonic echoes as a possible remnant.

Inner Planets — Mercury, Venus, Earth, and Mars — the four rocky planets within 1.6 AU of the Sun. Their positions are governed by Celtic Cross ratios referenced to Earth at 1.000 AU.

Outer Planets — Jupiter, Saturn, Uranus, Neptune, and Pluto — the five planets beyond the asteroid belt. Their positions are governed by multiples of *Harmonia*'s seed ratio, referenced to Jupiter at 5.204 AU.

Gas Giants — Jupiter (5.204 AU) and Saturn (9.559 AU), composed primarily of hydrogen and helium. In the model, Jupiter occupies the ratio $2AC$ and Saturn occupies $2AC^2$.

Ice Giants — Uranus (19.185 AU) and Neptune (30.156 AU), with interiors dominated by water, ammonia, and methane ices beneath thick hydrogen–helium atmospheres. In the model, Uranus occupies $4AC^2$ and Neptune occupies $2AC \cdot (B/D)$.

Dwarf Planet — A body orbiting the Sun with sufficient mass to be roughly spherical, but which has not cleared its orbital neighbourhood. Both Pluto (39.482 AU) and Ceres (2.77 AU) are classified as dwarf planets; both are included in the harmonic analysis.

Asteroid Belt — The region between Mars (~ 1.5 AU) and Jupiter (~ 5.2 AU) containing millions of rocky bodies. It occupies the zone where *Harmonia* would have resided, suggesting it may be the remnant debris of a disrupted body.

Ceres — The largest body in the asteroid belt and a dwarf planet, orbiting at 2.77 AU. Within 0.6% of the ratio C/A relative to *Harmonia* and within 0.5% of C relative to Mars — suggesting it may retain faint harmonic echoes of the original planetary structure.

Orbital Resonance — A gravitational relationship in which two orbiting bodies exert regular, periodic influences because their orbital periods are in a simple integer ratio. May have shaped the spacing pattern the Celtic Cross model describes.

Orbital Eccentricity — A measure of how much an orbit deviates from a perfect circle, from 0 (circular) to 1 (parabolic). The Celtic Cross model uses mean semi-major axes, unaffected by eccentricity, though eccentricity contributes to residual deviations from harmonic prediction.

Protoplanetary Disk — The rotating disk of gas and dust surrounding a young star from which planets form through accretion. The *Harmonia* hypothesis implies the Solar System's disk supported planet formation at ~ 2.14 AU before Jupiter's gravitational influence disrupted it.

Titius–Bode Law — An empirical rule (Titius 1766, Bode 1772) describing approximate planetary orbital distances by the formula $a = 0.4 + 0.3 \times 2^n$ AU. It predicted the asteroid belt and Uranus but fails significantly for Neptune and Pluto. The Celtic Cross model achieves a MAPE of 0.72% versus Titius–Bode's 2.46%, and has a geometric rather than empirical foundation.

Jupiter-Centric — A reference frame in which Jupiter's orbital radius (5.204 AU) serves as the unit of measurement for outer planet ratios: Saturn $\approx C \times a_J$, Uranus $\approx 2C \times a_J$, and so forth.

Harmonic Bridge — The Mars–*Harmonia*–Jupiter progression in which *Harmonia* connects the inner and outer harmonic systems. The ratio Mars/*Harmonia* $\approx 1/A$ and *Harmonia*/Jupiter $\approx D$ together complete the step missing from Kepler's model.

Harmonic Network — The full web of cross-planetary ratio relationships the model produces, encompassing not just consecutive planet pairs but non-adjacent pairs across the entire Solar System.

Ephemeris (pl. Ephemerides) — A dataset giving the calculated positions of celestial objects at regular time intervals. Modern ephemerides (NASA's DE series, France's INPOP, Russia's EPM) achieve positional accuracy of near metres over a century and are the observational foundation for the model's predictions.

Kirkwood Gaps — Regions in the asteroid belt with very few asteroids, caused by orbital resonances with Jupiter. Named after Daniel Kirkwood (1867). Mentioned as an analogue for the gravitational signature a harmonic node might leave in asteroid distributions.

Exoplanetary — Relating to planets orbiting stars other than the Sun. Future research directions include testing whether $\sqrt{2}$ harmonic patterns appear in multi-planet exosystems such as TRAPPIST-1 and Kepler-90.

N-body Simulation — A computational model calculating the gravitational interactions of N bodies simultaneously over time. Proposed as a method to test whether protoplanetary disks naturally evolve toward Celtic Cross harmonic spacing.

III. Historical & Cultural Terms

Music of the Spheres (*Musica Universalis*) — An ancient philosophical concept attributed to Pythagoras, proposing that the planets and stars move according to mathematical equations whose relationships correspond to musical intervals. Not literally audible, the harmony was understood as a structural property of the cosmos. Chapter 6 grounds this tradition in the model's semitone analysis, showing planetary positions match musical intervals within 0.21 semitones.

Sacred Geometry — The study of geometric forms and proportions considered to have cosmological significance across many ancient cultures. The Celtic Cross, the square-circle relationship, and ratios such as $\sqrt{2}$ and the Silver Ratio appear in traditions from Egypt, Scotland, India, and Mesoamerica. Used in this book to describe the geometric source of the harmonic constants, without implying supernatural causation.

Rosemarkie Stone — A carved Pictish stone from Rosemarkie, Scotland (c. 8th–9th century CE), cited as an example of the Celtic Cross construction with proportions clearly corresponding to the 3×3 grid and its derived ratios.

Royal Cubit — An ancient Egyptian unit of measurement (≈ 52.4 cm), defined as the length from the elbow to the tip of the extended middle finger plus a hand's width. Evidence that $\sqrt{2}$ proportions were encoded in practical measurement systems across ancient cultures.

Harmonices Mundi (*The Harmony of the World*) — A treatise by Johannes Kepler (1619) in which he proposed that the planets produce musical tones whose pitches depend on their orbital speeds. The most detailed pre-modern attempt to quantify the music of the spheres. The Celtic Cross model is argued to provide the geometric notation Kepler sought.

Mysterium Cosmographicum (*The Sacred Mystery of the Cosmos*) — Kepler's first major astronomical work (1596), proposing that planetary orbits are determined by the five Platonic solids nested inside one another. Though elegant, the model achieved accuracy of only several percent and could not accommodate Uranus and Neptune. The Celtic Cross model reduces Kepler's average error eightfold.

Platonic Solids — The five regular convex polyhedra: tetrahedron (4 faces), cube/hexahedron (6), octahedron (8), dodecahedron (12), and icosahedron (20). Kepler used these as the geometric key to planetary spacing. The cube and octahedron, as dual solids related through $\sqrt{2}$ and $1/\sqrt{2}$, contain the same proportional relationships as the Celtic Cross constants A and $1/A$.

Nested Solids — Kepler's geometric model in which each Platonic solid is inscribed within a sphere and circumscribed around another, and the ratio of the outer to inner sphere predicts the ratio of adjacent planetary orbits. Works approximately for the six planets Kepler knew but breaks down for the full Solar System.

Dual Solids — Two polyhedra whose vertices and faces correspond to each other. The cube and octahedron are duals; their proportional relationship produces exactly the ratios $\sqrt{2}$ and $1/\sqrt{2}$ — the harmonic core of the Celtic Cross model.

Cymatics — The study of visible patterns produced by vibrating media. When a surface covered with sand is vibrated at specific frequencies, the material organises into geometric patterns

— circles, stars, crosses, and spirals — determined by the frequency alone. Used as a physical analogue for the model's harmonic structuring principle.

Monochord — An ancient single-stringed instrument used by Pythagoras to demonstrate the relationship between string length and pitch. By dividing the string at simple ratios, the monochord showed that harmony arises from numerical proportion. Chapter 6 describes the Solar System as a "cosmic monochord" with planets sitting at harmonic nodes.

Mandala — A geometric or symbolic diagram used in Hindu and Buddhist traditions as a representation of the cosmos. Mentioned as a cultural parallel to the Celtic Cross's role as a geometric map of cosmic order. Used metaphorically.

Tritone — A musical interval spanning three whole tones (six semitones in standard tuning), equal to exactly half an octave. In Chapter 6's custom 10-semitone system, Mars's position corresponds approximately to a tritone above Earth (+6.08 semitones). Historically called *diabolus in musica* for its dissonant quality.

Cosmic Mediation — A metaphorical description of the Harmonic Triad: *Harmonia*'s empirical optimum lies between the algebraic Celtic Cross expression ($\sqrt{2}$) and the transcendental $\pi^{2/3}$, suggesting the Solar System balances discrete geometry with continuous curvature. Used metaphorically.

Harmonic Ladder — The complete sequence of planetary positions arranged by distance from the Sun, each rung corresponding to a specific Celtic Cross ratio. Visualised in the Appendix as an inner system ladder (referenced to Earth) and an outer system ladder (referenced to Jupiter).

Harmonic Ratio — A mathematical proportion expressing resonance or balance.

In this book, harmonic ratios are derived exclusively from the Celtic Cross geometry — specifically from $\sqrt{2}$ and its combinations — and govern the spacing of planetary orbits with a mean accuracy of 0.72%.

www.ingramcontent.com/pod-product-compliance
Lightning Source LLC
LaVergne TN
LVHW051129210926
848482LV00001B/24
9781837095209